Engineering Mechanics Lab Manual

The Authors

Dr. Ajay Kumar Gupta is presently working as Professor in Structural Engineering Department, M.B.M. Engineering College, Jai Narain Vyas University, Jodhpur. In his service span of 36 years he has also worked in the capacity of: Head of Structural Engineering Department, M.B.M. Engineering College, J.N.V. Univ., Jodhpur, Coordinator, State Resource Institute (Earthquake Risk Management-NPCBEERM, National Institute of Disaster Management), Director, State Retrofitting Clinic- Disaster Management & Relief, Chairman & Honorary Secretary of The Institution of Engineers (India) Jodhpur Local Center, Coordr. QIP (Poly) Cell, Coordr., State Technical Agency for Pradhan Mantri Gram Sadak Yojana (PMGSY), Vice President, MBMEC Alumni Association, Coordinator for the Entrepreneurship Dev. Cell, Convener Industry Institute Interaction Cell, President Vigyan Bharti, Jodhpur, Vice President, Vishawvidyalaya and Mahavidyalaya Shikshak Manch, Member-Shikshak Prakoshtha Rajasthan etc.

Published about One Hundred Technical papers in International/ National Conferences/Journals. Organised & participated in large numbers of Academics activities. Delivered approx Ninety Five Key note/ Lectures in numerous technical programmes at various places. Chaired number of sessions in seminars. As an expert in numbers of committees (e.g. syllabus, advisory, selection, technical), in (DTE, RUIDP, DMRC, JVVNL, Raj. Tech. Univ., Gujarat Technological University, RPSC, UPSC, AICTE, DST, Gujrat International Financial Technical City Technical Advisory Committee, Hazard Safety Cell- Rajasthan, Defense Lab-DRDO, Tech Education, Disaster Management & Relief, State Administration etc), MODROBS/ TAPTEC, DVP of INAE- AICTE, Examiner for number of M. E. thesis and Ph. D. thesis. Reviewed Books, Technical papers (American Concrete Institute-USA, Institution of Civil Engineers- UK), R&D Projects etc. Consultancy/ Non Destructive Testing assignments in & outside Rajasthan state. Supervising/ Supervised Seven Ph. D. and Supervised Thirty Five M.E. candidates for thesis/ seminar. Membership Tech. Bodies:- FIE, MISTE, MISRMTT, MICI, MISCMS, MIASE, MIACM, MISET, MINTACH

Er. Mohit Bhoot obtained his degree of Bachelor of Engineering in 'Civil Engineering' discipline from M.B.M. Engineering College, J.N.V. University, Jodhpur. He is currently working as an Assistant Engineer in Gujarat Water Supply & Sewerage Board (GWSSB) and is currently posted at Deesa (Guj). In the past, he has also worked in the capacity of 'Assistant Professor' and thus has teaching experience as well.

Engineering Mechanics Lab Manual

Dr. A.K. Gupta
Professor,
Structural Engineering Department,
M B M Engineering College, J N V University,
Jodhpur (Raj)

Mohit Bhoot
Assistant Engineer (Civil),
Public Health Sanitary Sub-Division,
GWSSB, Deesa (Guj.)

SCIENTIFIC
PUBLISHERS

Published by:

Scientific Publishers (India)
5 A, New Pali Road, P.O. Box 91
Jodhpur 342 001 (India)

E-mail: info@scientificpub.com
Website: www.scientificpub.com

Branch Office
Scientific Publishers (India)
4806/24, Ansari Road, Daryaganj
New Delhi - 110 002 (India)

ISBN: 978-81-7233-930-2

Typeset: Rajesh Ojha
Printed in India

Preface

Knowledge can be defined as familiarity with something, which can include some facts, information about a particular topic or subject, descriptions of experiments or study. It has been broadly classified into two types viz. Theoretical & Practical.

We cannot think of one being inferior to the other as both go hand in hand or side by side. `Temples' of learning the theoretical aspects of knowledge about any discipline are the classrooms & the lecture theatres while for the practical aspects, there are the laboratories. To catch up with the advances in the technology one must be well acquainted with the basic know how of the things. For the very same purpose this manual has been written.

Engineering Mechanics is a very crucial subject for almost all disciplines of engineering & sciences. The subject deals with the basic laws & fundamentals of the mechanics we encounter in our day to day, be it lever problem, phenomena related to friction, weight lifting machines, systems of pulleys & others.

The manual has been prepared in the form of a 'complete package' that includes, experiments which have been written very carefully meeting the standard adopted procedures, descriptive figures that aid the understanding, discussion section that intrigues the analytical & rational thinking, objective questions section & a wide reference list for detailed study. The language has been used keeping in view the wide readership which includes students, demonstrators, lecturers, field personnel & others. The selection of the experiments has been done very precisely, incorporating the very important ones from the subject.

Though ample efforts have been made to produce the manual with minimum flaws but they may have still crept in. We would be very gracious to the readers who would bring to our notice the errors in the text.

Last but not the least, we are also very thankful to Scientific Publishers for taking keen interest in the work & extending their support in bring out the manual in its present form.

Authors

Contents

EXPERIMENT NO. 1

To verify the Law of Parallelogram of forces & the Lami's Theorem

Lab Objective: To verify the Law of Parallelogram of forces & the Lami's Theorem.

Apparatus: Gravesand's apparatus, white drawing sheet, slotted weights, thread, small mirror strip, pencil, ruler, compass & protractor.

Theory: The apparatus essentially consists of a wooden board that is fixed on a wall. On the top corners of the board, two pulleys are mounted, one on each corner. It is assumed that the pulleys are frictionless. The arrangement of threads on the board is as shown in Fig. 1.1.

Parallelogram Law: The Law of Parallelogram states that "if two forces acting on a body at the same point are represented in magnitude and direction by the two adjacent sides of a parallelogram then, the resultant of the two forces is represented in magnitude and direction by the diagonal of the parallelogram at that point" (see Fig. 1.2).

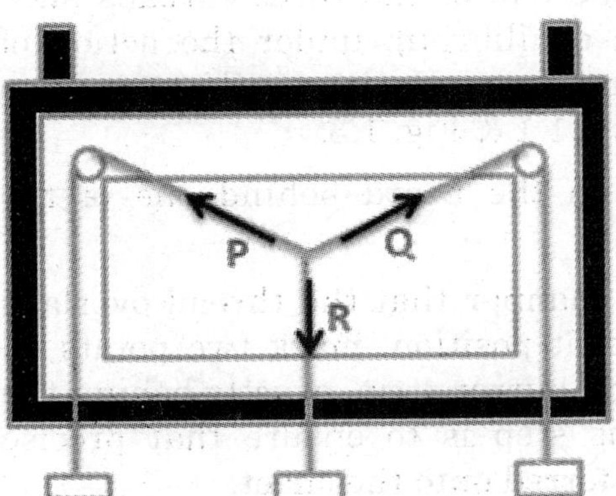

Fig. 1.1. Gravesand's Apparatus

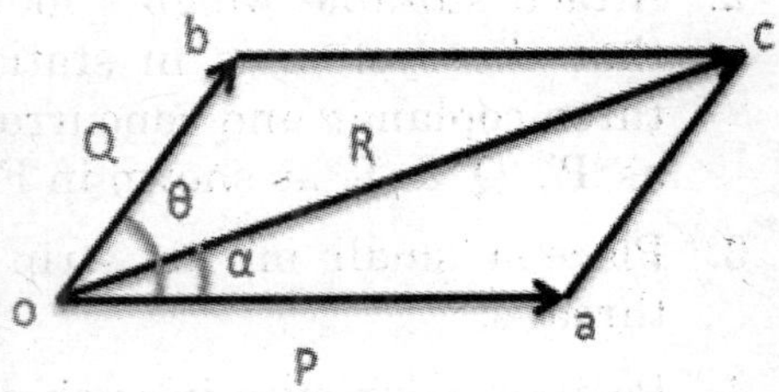

Fig. 1.2. Parallelogram of Forces

Lami's Theorem: If three concurrent forces, acting on a body are such that the system is in equilibrium then the forces follow the following relation.

$$\frac{P}{\sin\alpha} = \frac{Q}{\sin\beta} = \frac{R}{\sin\gamma}$$

In which 'P', 'Q' & 'R' are the three forces and 'γ', 'α' & 'β' are the angle between forces P & Q, Q & R and R & P respectively see Fig. 1.3.

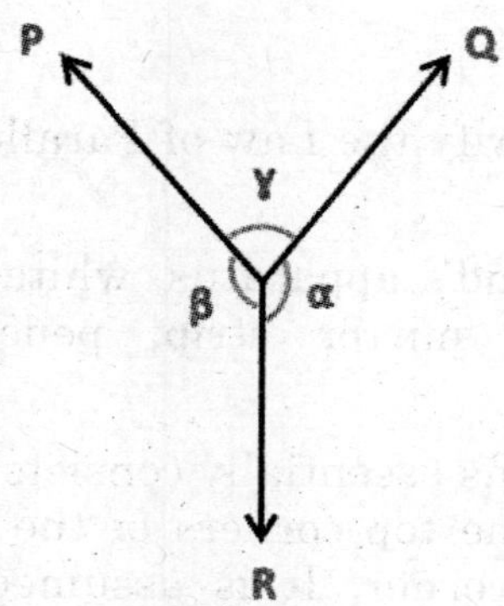

Fig. 1.3.

Procedure

To verify Parallelogram Law

1. Fix a white drawing sheet on the board of Gravesand's apparatus with the help of board pins or a tape.
2. Attach suitable weights at the end of the three threads such that the system is in static equilibrium under the action of three coplanar and concurrent forces. Designate these weights as 'P', 'Q' & 'R' as shown in Fig. 1.1 & Fig. 1.3.
3. Place a small mirror strip on the board behind one of the threads.
4. Position your eyes in such a manner that the thread overlaps its image in the mirror. In that position, mark two points on the paper, on either side of the mirror strip, exactly behind the thread. Main purpose of this step is to ensure that precise position of the thread is transferred onto the sheet.
5. Repeat the above two step for the other two threads also.

6. Remove the mirror and detach the sheet from the board. Join the points marked on the sheet to form three lines that meet at a common point, say O.
7. Now select a suitable scale to represent force in these threads on their respective lines. For this with O as centre cut arcs at points 'A', 'B' and 'C', on the lines in accordance with force in each thread (see Fig. 1.4).

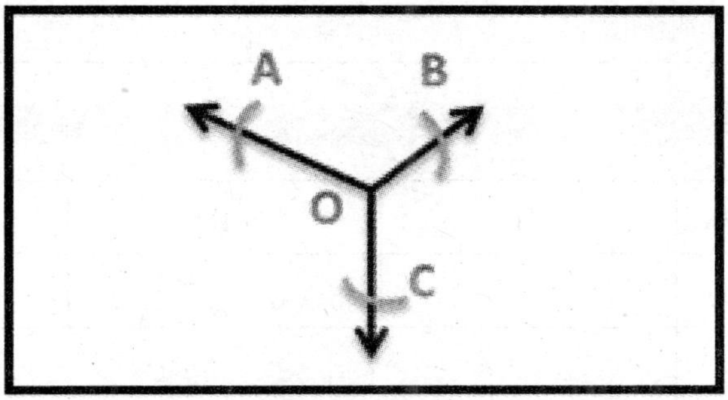

Fig. 1.4.

8. Taking OA & OB as adjacent sides, complete the parallelogram. Designate the last point as 'D'.
9. Next retrace the line OC in the reverse direction. Let this intersect AD or BD at point C'.
10. Measure the length of OC', convert it to weight in grams using the selected scale and record it as R'.

To verify Lami's Theorem

1. Measure the angle between three lines OA, OB & OC and designate them as α, β, & γ as shown in Fig. 1.5.

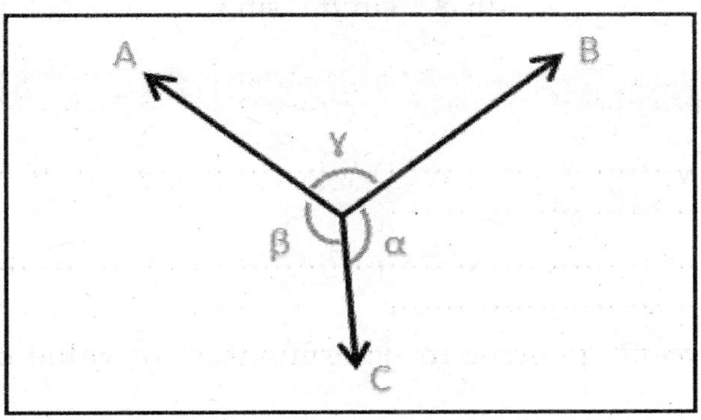

Fig. 1.5.

Observations:

Scale adopted-

Table 1

Serial No.	P (g)	Q (g)	R (g)	R' (g)

Table 2

Serial No.	α	β	γ

Calculations:

Error in value of resultant of forces P & Q = R – R'

= .... g.

Percentage Error $= \frac{R - R'}{R} \times 100$

= .... %.

To verify Lami's Theorem, substitute values in the following formula from Table 1 & Table 2.

$$\frac{P}{\sin \alpha} = \frac{Q}{\sin \beta} = \frac{R}{\sin \gamma}$$

Results:

1. ....

2. ....

3. The percentage error in determination of value of resultantis ....

Discussions:

1. What are the major sources of error in this experiment?
2. Why is it necessary to bring eyes in proper position before marking thread image on sheet?
3. What are the two necessary conditions checked before using Lami's Theorem in a particular system?

Space for Answers:

EXPERIMENT NO. 2

To verify Polygon law of Forces

Lab Objective: To verify the Polygon Law of forces.

Apparatus: Gravesand's apparatus, threads, white drawing sheet, board pins or tape, mirror strip, ruler, compass, protractor & slotted weights.

Theory: The apparatus essentially consists of a wooden board that is fixed on a wall. On the top & bottom corners of the board, 4 pulleys are mounted, one on each corner. It is assumed that the pulleys are friction-less. The arrangement of the threads on the board is as shown in Fig 2.1.

Polygon Law: The Polygon Law of forces states that "if a number of forces acting on a body are represented in magnitude & direction by the sides of a polygon taken in the 'same order', then the resultant of all such forces is represented in magnitude & direction by the closing side of the polygon taken in 'opposite direction'".

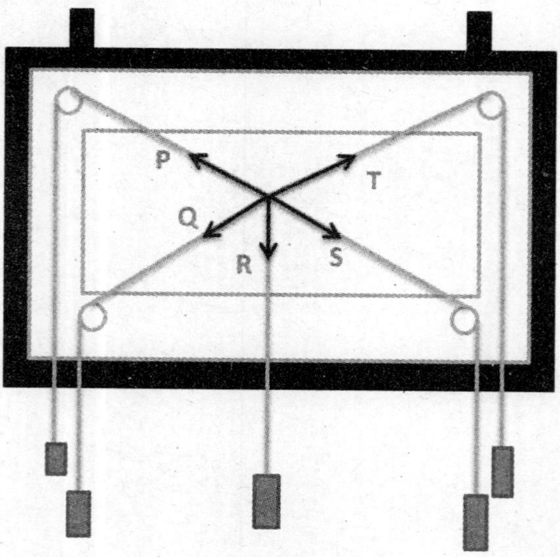

Fig 2.1. Gravesand's apparatus

Procedure

1. Fix a white drawing sheet on the board of Gravesand's apparatus with the help of board pins or tape.
2. Attach suitable weights at the ends of all these threads such that the system is in static equilibrium and the common knot is approximately in the centre of sheet. Designate these weights as 'P', 'Q', 'R', 'S', & 'T', as shown in fig 2.1.
3. Place a small mirror strip under one of the thread and position your eyes such that the thread and its image in the mirror overlap each other. Now in this position, mark on the drawing sheet the extremities of the image on either side of the mirror. Main idea behind doing this is to transfer the precise position of thread on the sheet.
4. Repeat step 3 for the other threads also.
5. Now remove the sheet from the apparatus. Using a ruler, join points marked for each thread to form five lines intersecting at one point (see fig 2.2).
6. Adopt a suitable scale to represent the force in each thread.
7. Select a clockwise working direction (students may choose anticlockwise as well) to draw the vector force diagram.

TO DRAW POLYGON:

1. Draw 'ab' parallel to OP and having length as per adopted scale. Mark the arrow properly to represent the direction of force.
2. Next, draw 'bc' parallel to OQ and having length as per adopted scale. Again proper marking of arrows is important. Similarly, draw 'cd' & 'de' to represent forces OR & OS respectively.
3. As per Polygon Law of forces, in this case, the last side must be the resultant of forces taken opposite in direction. Therefore, complete the polygon by joining 'a' & 'e' and mark the arrow in the reverse direction. 'ae' represents force T'(see fig 2.3).
4. Measure 'ae' and determine the force T' using the adopted scale.

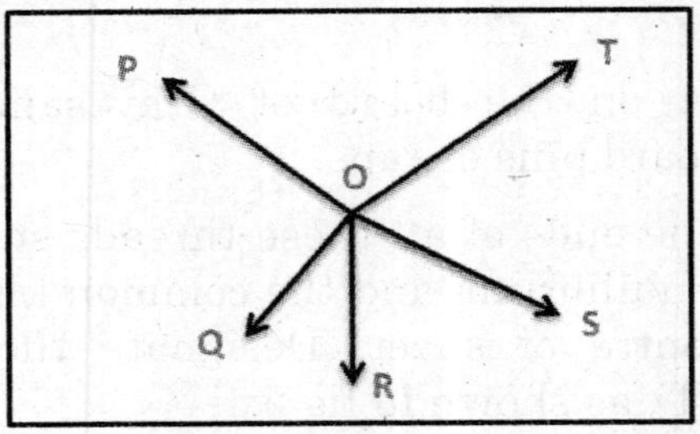

Fig. 2.2.

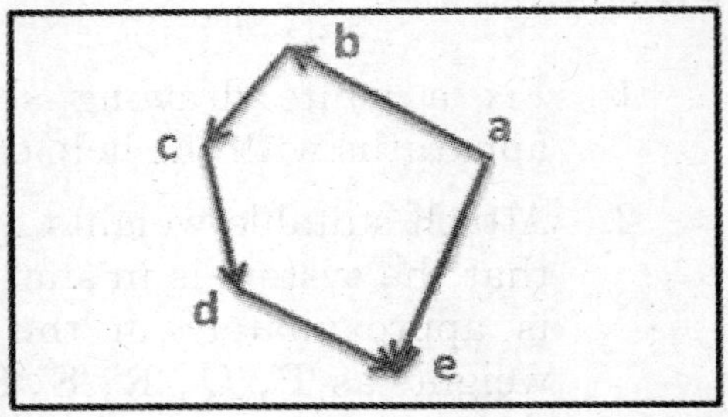

Fig. 2.3.

Observations:

Scale –

Table

Serial No.	P (g)	Q (g)	R (g)	S (g)	T (g)	T' (g)

Calculations:

Error in value of resultant force $= T - T'$

$= \ldots\ldots\ldots\ldots$ g.

Percentage Error $= \dfrac{T - T'}{T} \times 100$

$= \ldots\ldots\ldots\ldots$ %.

Results:

1. ....

2. Percentage error in determination of value of resultant is ....

Discussions:

1. What are the sources of error in the experiment?
2. Polygon Law is an extension of Parallelogram Law. Comment.

Space for Answers:

EXPERIMENT NO. 3

To determine the forces in the members of a Jib Crane

Lab Objective: To determine the forces in the members of a Jib Crane.

Apparatus: Jib Crane setup, weights, metre scale, slotted weights, threads etc.

Theory: Crane is basically a machine used for the purpose of lifting heavy weights and transporting them from one place to another. Jib Crane is a type of such crane. The Jib Crane setup that we encounter in lab is generally a miniature form.

Triangle Law: The experiment is basically performed for verification of Triangle Law of forces that states "if two forces acting simultaneously on a body are represented in magnitude & direction by the two adjacent sides of a triangle taken in same order then their resultant may be represented in magnitude & direction by the third side taken in opposite order" see (fig 3.1).

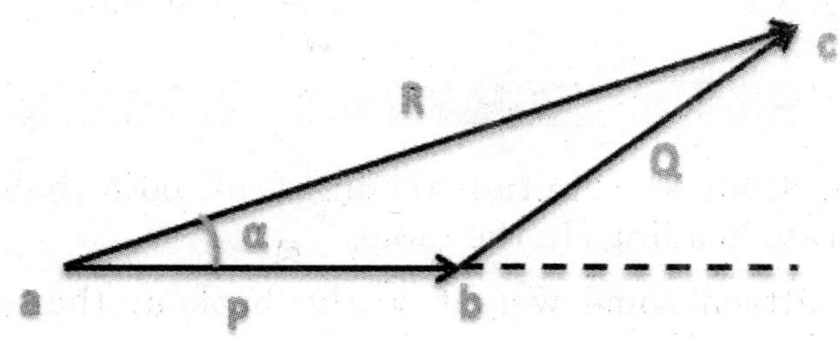

Fig 3.1

The Jib Crane apparatus varies from laboratory to laboratory. Its either a wall mounted setup or the portable one. The three major components of Jib Crane are

(a) A vertical tubular rod that carries hooks at different heights.

(b) A tie rod, whose one end is connected to the third component called the Jib, while the other end can be attached to any of the hooks of the vertical tubular rod. To the tie rod, is attached a spring balance that measures the tension in the tie rod.

(c) A Jib, which is an inclined wooden rod. One of its end is connected to the tie rod, while on the other is provided a compression balance for determining the compression force in the Jib. The end which is connected to tie rod is also provided with a hook and a chain to which some weight may be suspended.

In case of a portable Jib Crane apparatus, the vertical rod is mounted on a heavy base that provides stability and rigidity to the apparatus. See fig 3.2 and fig 3.3.

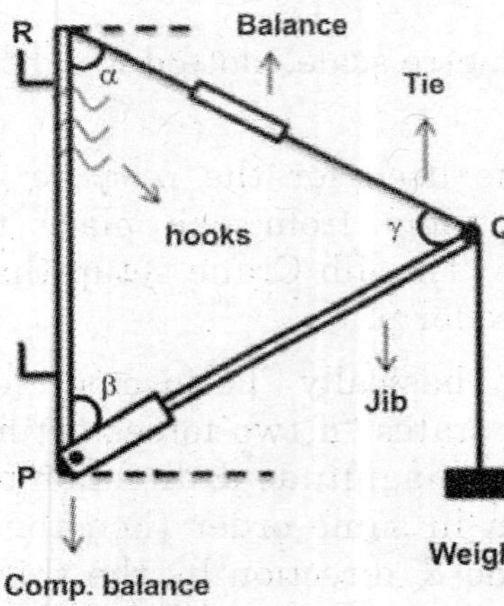

Fig. 3.2.

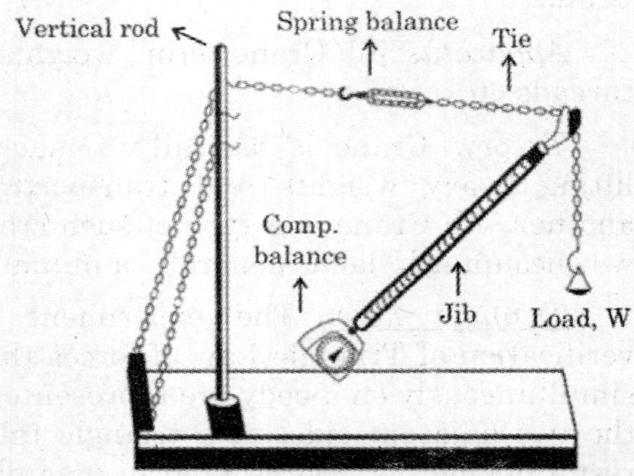

Fig. 3.3.

Procedure

1. Note down the initial readings of both balances under separate headings in the table.
2. Now attach some weight to the hook at the end of chain, record it as 'W'.
3. Record the final readings in the spring & the compression balance.
4. With the help of a thread & metre scale measure the length of the Jib, the tie & the vertical rod.

Observations:

Length of the Jib, PQ = ... cm.

Length of the post, RP = ... cm.

Length of the tie, QR = ... cm.

Table

Serial No.	Reading of Spring Balance, T (g)		Reading of Compression Balance, C (g)		Weight suspended, W (g)	Percentage Error in value of force in tie (%)	Percentage Error in value of force in jib (%)
	Initial	Final	Initial	Final			
1.							
2.							

Calculations:

We need to determine value of angles 'α' &'β'. For this, consider fig 3.4.

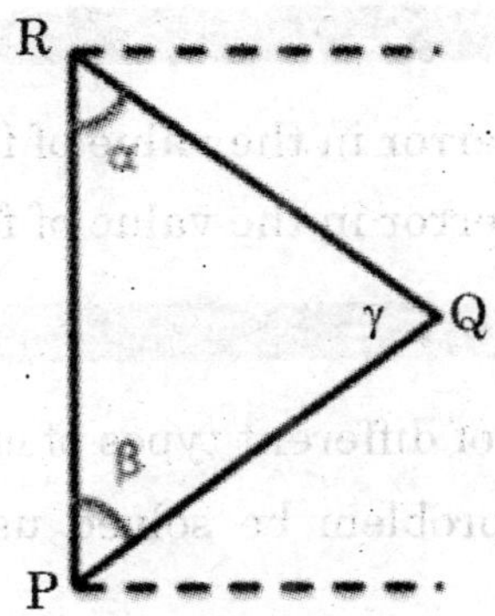

Fig. 3.4.

In triangle PQR, using cosine formula,

$$\cos \alpha = \frac{RP + QR^2 - PQ^2}{2 \times RP \times QR}$$

&

$$\cos \beta = \frac{RP^2 + PQ^2 - QR^2}{2 \times RP \times PQ}$$

Also, $\alpha + \beta + \gamma = 180^\circ$

Or $\gamma = 180^\circ - (\alpha + \beta)$

Thus,

The value of α = ______.

The value of β = ______.

The value of γ = ______.

Now using the Sine law,

$$\frac{W}{\sin\gamma} = \frac{C'}{\sin\beta} = \frac{T'}{\sin\alpha}$$

Where, C' & T' are the calculated values of the force in Jib & tie respectively.

Calculate the values of C' & T'. Also compute & record the percentage error in the values of force in the jib & tie in the respective columns in the table.

$$\text{Percentage error} = \frac{\text{Observed value} - \text{Calculated value}}{\text{Observed value}} \times 100$$

Results

(1) The percentage error in the value of force in jib = ... %.

(2) The percentage error in the value of force in tie = ... %.

Discussion:

(1) Give the names of different types of cranes.

(2) Can the same problem be solved using Lami's Theorem? If yes, how?

Space for Answers:

EXPERIMENT No. 4

To determine the Support Reactions of a Simply Supported Beam

Lab Objectives: To determine the Reactions at the support of a Simply Supported Beam.

Apparatus: A graduated beam (preferably wooden), slotted weights, compression balances, sliding hangers, etc.

Theory: The whole apparatus is set upon a wooden or metal board that serves the purpose of giving stability to the whole setup. The beam has a uniform cross-section throughout and is mostly 1 metre in length. The beam is graduated for the precise measurement of the distances of the weights from the supports. Also, grooves are cut on the beam throughout, for easy application of loads via hangers. Two compression balances are also installed that directly give the value of support reactions in 'g' or 'kg' (see fig 4.1).

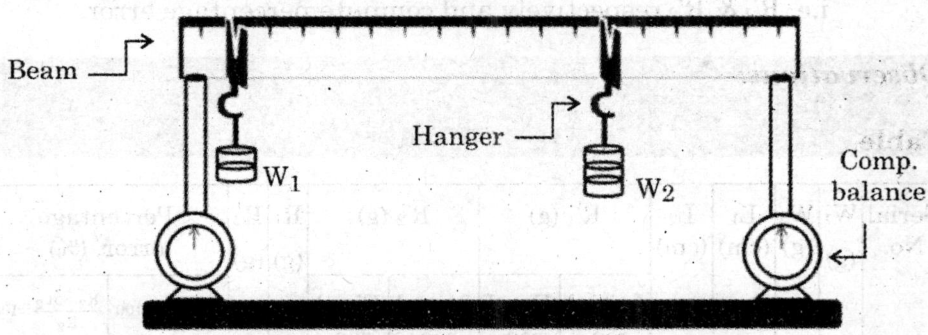

Fig. 4.1

In the experiment, reactions at the support are determined using the 'Principle of Moments', according to which if a number of coplanar forces are acting on a body in equilibrium then the algebraic sum of moment of all such forces taken about any point in the plane is zero. Thus,

$$\Sigma M = 0$$

Also, we know that for a body in equilibrium, the algebraic sum of all the forces in horizontal direction & algebraic sum of all the forces in vertical direction is equal to zero. Thus,

$$\sum H = 0 \quad \& \quad \sum V = 0$$

Procedure:

1. Record the initial reading of both the balances.
2. Now suspend some weight on the beam through the hangers at different points on the beam.
3. Let the weights be designated 'W_1' & 'W_2' and the distance of these weights from the left hand support be 'L_1' & 'L_2', respectively.
4. Record the value of W_1, W_2, L_1 & L_2
5. Also record the final reading of both the compression balances as R'_1 & R'_2.
6. Now, repeat the above steps with different sets of loading applied at different points along the beam. Take minimum three such sets of readings.
7. Analytically determine the values of support reactions R_1 & R_2 and compare them with the experimentally obtained values i.e. R'_1 & R'_2 respectively and compute percentage error.

Observations:

Table

Serial No.	W_1 (g)	W_2 (g)	L_1 (cm)	L_2 (cm)	R'_1 (g)		R'_2 (g)		R_1 (g)	R_2 (g)	Percentage error, (%)	
					Initial reading	Final reading	Initial reading	Final reading			$\frac{R_1 - R'_1}{R_1} \times 100$	$\frac{R_2 - R'_2}{R_2} \times 100$
1.												
2.												
3.												
4.												

Calculations:

Consider the beam shown below

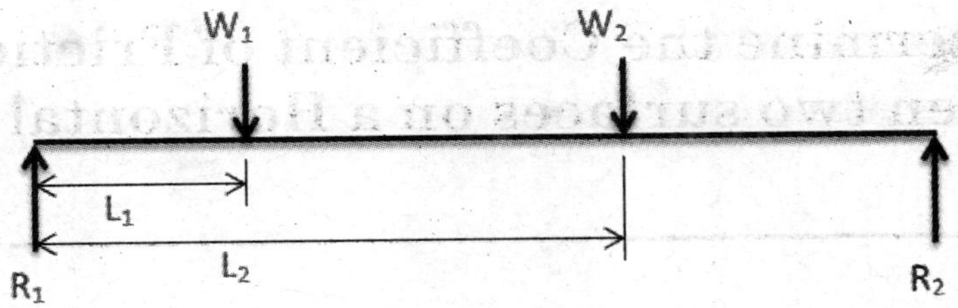

Fig. 4.2

Here,

W_1 & W_2 are the applied loads on the beam of length L, at distances L_1& L_2, respectively.

R_1 & R_2 are the reactions at support, which are determined as follows

Taking moment of all forces about R_1,

$$R_2 \times L - W_2 \times L_2 - W_1 \times L_1 = 0$$

Or $$R_2 = \frac{W_1 \times L_1 + W_2 \times L_2}{L}$$

Since $\sum V = 0$

$$R_1 + R_2 = W_1 + W_2$$

Or $$R_1 = W_1 + W_2 - R_2$$

Results:

1. Percentage error in determination of R_1 = ... %.
2. Percentage error in determination of R_2 = ... %.

Discussions:

1. How many types of supports are there?
2. What do you understand by a simply supported beam?
3. Do you think if the loads were applied at some inclination to the beam axis, the results would have been different?

Space for Answers:

EXPERIMENT No. 5

To determine the Coefficient of Friction between two surfaces on a Horizontal Plane

Lab Objectives: To determine the coefficient of friction between two surfaces on a horizontal plane.

Apparatus: Horizontal surface board with pulley, slotted weights, woollen cloth strip, wooden plank, thin iron/aluminium/steel sheet, small pan, thread or rope, wooden/iron box.

Theory: When a body slides or tries to slide over another body (assumed stationary), then a resistive force comes into action at the surface of contact that tries to oppose this relative motion This resistive force is termed as friction or frictional force. The friction is a self-adjusting force i.e. it varies with the change in the force causing the motion. But if the force causing motion exceeds a certain limit the friction becomes constant.

In the present experiment we have a horizontal surface board at one edge of which, a small pulley is installed. A small wooden or iron box is taken which has a small hook fixed to one of its side. A thread is entangled into this hook and then taken over the pulley. To the free end of this thread a small pan is fastened.

Now, consider the motion of the box over the horizontal board,

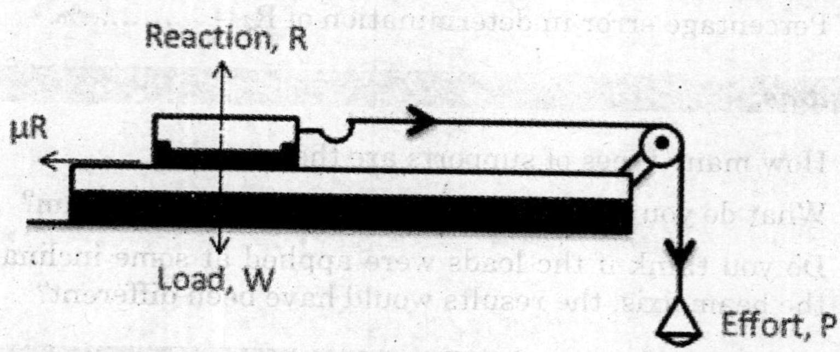

Fig. 5.1.

Under equilibrium condition,

$$\sum V = 0$$

$$\therefore R = W \quad \text{... eq. 1}$$

$$\sum H = 0$$

$$\therefore \mu R = P \quad \text{... eq. 2}$$

Or, $\mu R = P$

Or, $\mu(W) = P$ ... eq. 3

Procedure:

1. Determine and note the weight of empty pan and wooden box. Let these be 'p' and 'w' respectively.
2. Setup the board in horizontal position with rope or thread connecting the wooden from box to the pan, while going over the pulley.
3. Draw the box just near the far edge of wooden board.
4. Place a suitable slotted weight (say 200 gm) in the wooden/ iron box. Record this 'total weight' as 'W'.
5. Next put some suitable weight in the pan and gently tap the surface of the board such that the box 'just' starts sliding. If no sliding occurs, gradually increase the weight in the pan and continue tapping.
6. The weight for which the box 'just' begins to slide is recorded as 'P'.
7. Repeat the above procedure to obtain minimum 3 different set of observations
8. The entire procedure can be repeated to determine the observations for different pair of material (say wood and aluminium/iron/wool).

Obseravation:

Weight of empty box, w = ... g.

Weight of empty pan, p = ... g.

Table:

Serial no.	Load, W (g) (including w)	Effort, P (g) (including p)	Coefficient of friction μ	Average value of μ
1.				
2.				
3.				
4.				

Calculations:

Coefficient of friction, $\mu = \frac{P}{W}$

Graph:

Results:

1. The average value of coefficient of friction between .... &............................ is ....

Discussions:

1. "Friction is a necessary evil". Justify.
2. Differentiate between coefficient of static friction and kinetic friction.
3. Enumerate the possible errors that might creep in the above experiment.

Space for Answers:

EXPERIMENT No. 6

To determine the Co-efficient of Friction between two surfaces on an Inclined Plane

Lab Objectives: To determine the coefficient of friction between two surfaces on an inclined plane.

Apparatus: Adjustable surface board with pulley, slotted weights, woollen cloth strip, wooden plank, thin iron/aluminium/steel sheet, small pan, thread or rope, wooden/iron box.

Theory: Friction as we already know is a self-adjusting resistive force that comes into play at the surface of contact when one body moves or tries to move over another body. Initially when the force is not sufficient for relative motion between the bodies, the frictional force is same as the applied force but opposite in direction. But as the force is gradually increased, there comes a state at which the body just begins to move, at that state maximum frictional force is developed and is known as the limiting force of friction.

In the present experiment we have an inclined board on the surface of which is placed an open box of wood or iron. To one side of this box a hook is fixed. A rope or a string is attached to this hook and is taken over a pulley up the plane. The free end of this string is tied to a small pan.

Now, consider the motion of the box up the inclined plane,

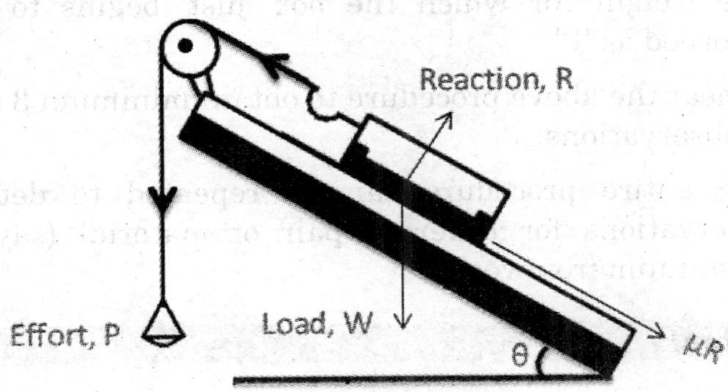

Fig. 6.1.

Under the conditions of limiting equilibrium,

Resolving forces in the direction parallel to the plane and summing,

$$W \sin\theta + \mu R = P \quad \text{... eq. 1}$$

Resolving forces in the direction perpendicular to the plane and summing,

$$R = W \cos\theta \quad \text{... eq. 2}$$

Using eq. 1 and eq. 2 we get,

$$P = W \sin\theta + \mu W \cos\theta$$

Or

$$\mu = \frac{P - W\sin\theta}{W\cos\theta} \quad \text{... eq. 3}$$

Procedure:

1. Determine and note the weight of empty pan and the box. Let these be 'p' and 'w' respectively.
2. Set up the board in a suitable inclined position and determine its angle of inclination with horizontal. Record it as 'θ'.
3. Fasten the rope or thread connecting the wooden box to the pan, while it goes over the pulley.
4. Draw the box just near the hinges.
5. Place a suitable slotted weight (say 200 gm) in the wooden box. Record this 'total weight' as 'W'.
6. Next put some suitable weight in the pan such that the box 'just' starts sliding up. If no sliding occurs, gradually increase the weight in the pan.
7. The weight for which the box 'just' begins to slide up is recorded as 'P'.
8. Repeat the above procedure to obtain minimum 3 different set of observations.
9. The entire procedure can be repeated to determine the observations for different pair of material (say wood and aluminium/iron/wool).

Obseravation:

Weight of empty box, w = ... g.

Weight of empty pan, p = ... g.

Angle of inclination of board, θ = ...

Table:

Serial No.	Load, W (g) (including w)	Effort, P (g) (including p)	Coefficient of friction μ	Average value of μ
1.				
2.				
3.				
4.				

Calculations:

$$\text{Coefficient of friction, } \mu = \frac{P - W\sin\theta}{W\cos\theta}$$

Graph:

Results:

1. The average value of coefficient of friction between .... & .... is .... on inclined plane with angle of inclination with horizontal equal to ....

Discussions:

1. "Friction is a necessary evil". Justify.
2. Differentiate between coefficient of static friction and kinetic friction.
3. Enumerate the possible errors that might creep in the above experiment.

Space for Answers:

EXPERIMENT No. 7

To verify the principle of moments using Bell Crank Lever

Lab Objectives: To verify the principle of moments using Bell Crank lever.

Apparatus: Bell crank lever apparatus, hangers & hanger weights, good quality ruler etc.

Theory: Levers are the most primitive invention of mankind for the purpose of lifting loads. It is a rod or a beam that may be either straight or curved, pivoted at a point. This point is known as the Fulcrum. Depending upon the position of the load, the fulcrum & the effort along the rod, there are three types of levers namely, Simple levers, Crank levers & Compound levers.

Principle of Moments: If a body is acted upon by a number of co-planar forces, then for equilibrium of the body, the algebraic sum of moment of all these forces about any point lying in the plane is zero. Thus,

$$\sum M = 0$$

Certain terms in context of levers to which the students should be well acquainted are,

1. Resistance arm – It is the portion of rod between the fulcrum and the point of application of load.
2. Effort arm - It is the portion of rod between the fulcrum and the point of application of effort.
3. Leverage – It is defined as the ratio of the effort arm & the resistance arm. Sometimes, it is also reported as the ratio of the load & the effort.

The apparatus consists of a 90° bent lever with one arm shorter than the other. The longer arm is graduated & grooved slightly for the ease of distance measurement & placing the hangers respectively. The lever is pivoted on a platform at the bent, forming the fulcrum. The longer side is kept horizontal and thus the shorter side stands vertical.

On the platform at some distance away from the shorter arm a vertical post is fixed parallel to the former. This post is connected to the shorter arm of lever through a spring balance and chain setup. Here the effort is measured. A thin upright rod is provided as reference near the short arm to bring it to precise vertical position. Fig 7.1 shows the schematic diagram of the apparatus while the fig 7.2 shows the line diagram.

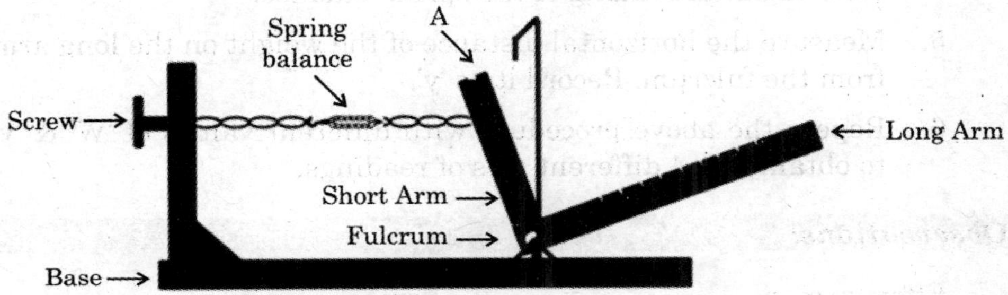

Fig. 7.1

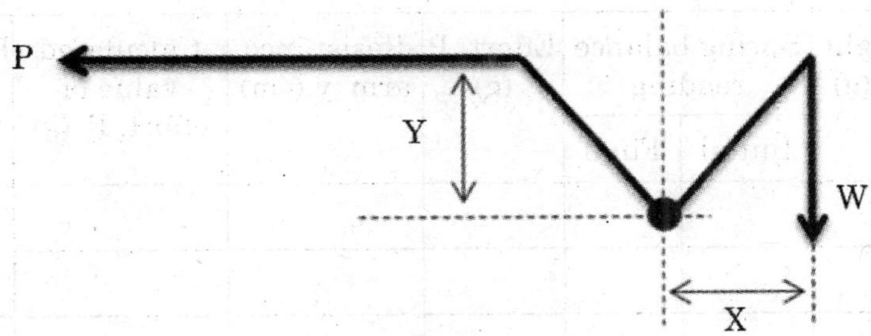

Fig. 7.2

Considering Fig 7.2 and taking algebraic sum of moment of all forces about the fulcrum, we have,

$$P.y - W.x = 0 \quad \text{...eq. 1}$$

Or, $$P = \frac{Wx}{y} \quad \text{...eq. 2}$$

Or, $$\frac{W}{P} = \frac{y}{x} = Leverage \quad \text{...eq. 3}$$

Procedure:

1. With the help of the screw bring the short arm to its vertical position. The short bar is truly vertical when the point A is just over the groove on the top of short arm.

2. Note the initial reading of the spring balance. Also measure the vertical distance of the point iron where the chain is hooked on the short arm, and the fulcrum. Record it as 'x'.
3. Suspend some weight with the help of hanger on the horizontal long arm. Record this weight as 'W'.
4. Again adjust the screw so that the short arm is again vertical. Note the final reading of the spring balance.
5. Measure the horizontal distance of the weight on the long arm from the fulcrum. Record it as 'y'.
6. Repeat the above procedure with different values of 'W' & 'y' to obtain 3 or 4 different sets of readings.

Observations:

Effort arm, x = ... cm.

Table

Serial No.	Weight, W (g)	Spring balance reading		Effort, P (g)	Resistance arm, y (cm)	Calculated value of effort, P' (g)	Percentage Error (%)
		Initial	Final				
1.							
2.							
3.							
4.							

Calculations:

1. Calculate the value of P' for each set of reading by putting values of 'W', 'x'& 'y' in

$$P' = \frac{Wx}{y}$$

2. Also calculate the error by using,

$$\text{Percentage Error} = \frac{P - P'}{P} \times 100\%$$

Results:

1. Within experimental errors the principle of moments stands ....

Discussions:

1. What are the different types of levers? Also give example of each type.
2. Name a few applications of Crank levers.
3. What advantage does Crank lever have over the Simple levers?

Space for Answers:

EXPERIMENT No. 8

To verify the principle of moments using Compound lever

Lab Objectives: To verify the principle of moments using Compound lever.

Apparatus: Compound lever apparatus, hangers & hanger weights, good quality ruler etc.

Theory: A compound lever is a combination of two simple levers, one of first class and another of second class. Such a system yields greater leverage and mechanical advantage. Thus it can be used to lift heavier loads with minimum effort. The experiment is performed to verify the principle of moments.

Principle of Moments: If a body is acted upon by a number of co-planar forces, then for equilibrium of the body, the algebraic sum of moment of all these forces about any point lying in the plane is zero. Thus,

$$\sum M = 0$$

The apparatus consists of simple levers AF_1B and F_2EG with their fulcrum at F_1 and F_2 respectively. The lever AF_1B is class I simple lever while the lever F_2EG is class II simple lever. Both the levers are pin-joined through a rod at points B and G (see fig 8.1). The load is applied at point E while the effort to lift this weight is applied at point A.

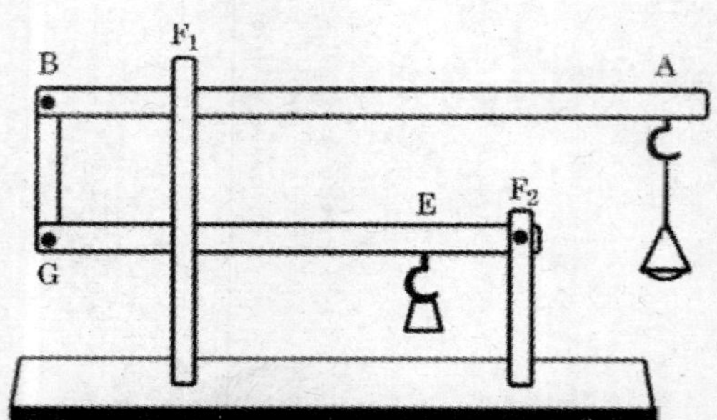

Fig. 8.1.

Now consider the equilibrium of the system (fig 8.2) and let,

W = load to be lifted at point E.

P = effort applied at point A.

R = reaction in member GB.

a = distance BF_1.

b = distance AF_1.

c = distance GF_2.

d = distance EF_2.

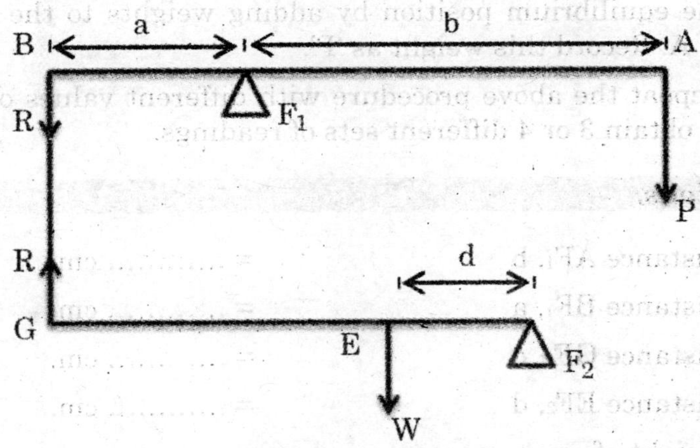

Fig. 8.2.

In Simple lever, AF_1B taking moment of forces about fulcrum F_1,

$$R.\,a = P\,.\,b$$

Or $$R = \frac{Pb}{a}$$ ...eq. 1

In Simple lever, F_2EG taking moment of forces about fulcrum F_2,

$$R.\,c = W.\,d$$

Or $$R = \frac{wd}{c}$$ ...eq. 2

Comparing the value of R from eq. 1 and eq. 2, we have,

$$\frac{Pb}{a} = \frac{Wd}{c}$$

Or $$P = W.\frac{ad}{bc}$$ ...eq. 3

Procedure:

1. Attach some load to the hanger at E and bring the lever AF_1B in horizontal position by putting some weight in the effort pan at A. This is done to bring the levers in horizontal position & parallel to each other.
2. Measure accurately and record the distances 'a', 'b', 'c' &'d'.
3. Suspend some weight with the help of hanger at E. Record this weight as 'W'.
4. Determine the weight of empty pan. Record it as 'w'. Restore the equilibrium position by adding weights to the effort pan at A. Record this weight as 'P'.
5. Repeat the above procedure with different values of 'W' & 'P' to obtain 3 or 4 different sets of readings.

Observations:

Distance AF_1, b = ... cm.

Distance BF_1, a = ... cm.

Distance GF_2, c = ... cm.

Distance EF_2, d = ... cm.

Weight of empty pan, w = ... g.

Table

Serial No.	Load, W (g)	Effort, P (g) (including 'w')	Calculated value of effort, P' (g)	Percentage Error (%)
1.				
2.				
3.				
4.				

Calculations:

1. Calculate the value of 'P' for each set of reading by putting values of 'W', 'a', 'b', 'c' & 'd' in

$$P' = W \cdot \frac{ad}{bc}$$

2. Also calculate the error by using,

$$\text{Percentage Error} = \frac{P - P'}{P} \times 100\%$$

Results:

1. Within experimental errors the principle of moments stands ....

Discussions:

1. What are the different types of levers? Also give example of each type.
2. Name a few applications of Crank levers.
3. What advantage does Crank lever have over the Simple levers?
4. Suggest some possible uses of compound lever system.

Space for Answers:

EXPERIMENT No. 9

To determine the Mechanical Advantage, Velocity Ratio & Efficiency of First system of Pulleys

Lab Objectives: To determine the Mechanical Advantage, Velocity Ratio & Efficiency of First System of Pulleys.

Apparatus: First system of Pulley apparatus, small pan, threads or strings, slotted weights, small weight box, metre rod etc.

Theory: The arrangement of pulleys is as shown in fig 9.1. There are 3 movable pulleys and 1 fixed pulley. The fixed pulley is provided so that application of effort is easy. One end of each of the three threads is fixed to a wooden or metallic rod or frame. The other end of each of these threads after going round its respective pulley is hooked to the next higher pulley in order. To the free end of the thread going round the fixed pulley a small pan is fastened for application of effort. The load to be lifted is hooked through a separate thread from the lowest pulley. This system of pulley is also known as 'separate string system'.

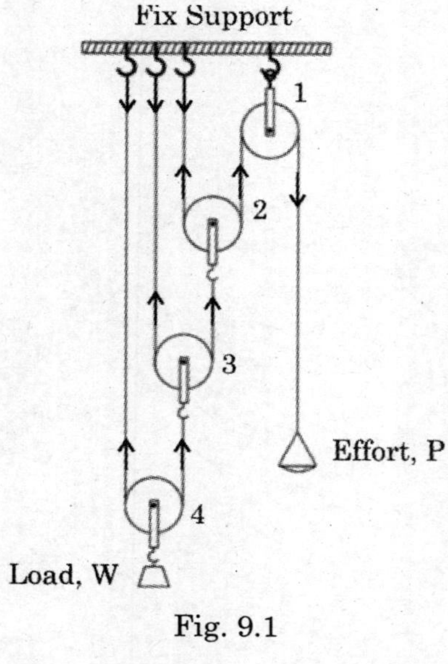

Fig. 9.1

Force Distribution:

In both strings of Pulley 1 = P

In both strings of Pulley 2 = P

In both strings of Pulley 3 = 2P

In both strings of Pulley 4 = 4P

Wherein, Pulleys 2, 3 & 4 being the movable ones while the Pulley 1 is fixed.

Now,

a. Mechanical Advantage – It is defined as the ratio of Load lifted and the Effort applied.

$$M.A. = \frac{W}{P}$$

b. Velocity Ratio – It is defined as the ratio of the distance moved by the Effort and the Load. Here,

If the effort pan moves a distance say x, then

Distance moved by the pulley 2 = $x/2 = \frac{x}{2^1}$

Distance moved by the pulley 3 = $\frac{1}{2}\times\frac{x}{2} = x/4 = \frac{x}{2^2}$

Distance moved by the pulley 4 = $\frac{1}{2}\times\frac{x}{4} = x/8 = \frac{x}{2^3}$

Therefore, distance moved by the load = $\frac{x}{2^3}$

Working on same lines, if there were 'n' movable pulleys, then the distance by which the load moves =

$$V.R. = \frac{\text{distance moved by effort}}{\text{distance moved by the load}}$$

$$= \frac{x}{(x/2^n)} = 2^n$$

c. Efficiency – It is defined as the ratio of the Mechanical Advantage and the Velocity Ratio. It is expressed as a percentage.

$$\eta = \frac{M.A.}{V.R.} \times 100$$

Procedure:

1. Make sure that the apparatus is properly functioning with minimum resistance or friction.
2. Hang some weight to the lowest movable pulley. Record it as Load, W.
3. Determine the weight of empty effort pan. Let this be 'w'. Add some weight to the effort pan. Keeping on adding more weight until the load, W just starts lifting. Record this value of Effort as 'P'.
4. Repeat the steps 2 & 3 two more times to obtain total 3 sets of readings.

For Velocity Ratio:

1. Determine the initial position of the effort pan and the load with respect to the ground with the help of metre rod. Let these be 'y_1' and 'x_1' respectively.
2. Now, pull the effort pan down a certain distance and measure the height of the effort pan and the load with respect to ground. Let these be 'y_2' and 'x_2' respectively.

Observations:

Number of movable pulleys in the system, n = ...

Weight of the empty pan, w = ... g.

Initial position of effort pan, y_1 = ... cm.

Final position of effort pan, y_2 = ... cm.

Initial position of Load, x_1 = ... cm.

Final position of Load, x_2 = ... cm.

Table

Serial No.	Load, W (g)	Effort, P (g) (including 'w')	M.A.	V.R.	Efficiency, η (%)
1.					
2.					
3.					

Calculations:

$$\text{M.A.} = \frac{W}{P}$$

$$\text{V.R.} = \frac{y_2 - y_1}{x_2 - x_1}$$

Verify V.R. = 2^n

$$\text{Efficiency, } \eta = \frac{M.A.}{V.R.} \times 100$$

Results:

For the given 1st system of pulley,

1. Mechanical Advantage = ...
2. Velocity ratio = ...
3. Efficiency = ... %.

Discussions:

1. List a few uses of pulley in daily life.
2. Can the M.A. be increased in this system by increasing the number of pulleys?
3. How can the efficiency be increased in this system of pulleys?

Space for Answers:

EXPERIMENT No. 10

To determine the Mechanical Advantage, Velocity Ratio & Efficiency of Second system of Pulleys

Lab Objectives: To determine the Mechanical Advantage, Velocity Ratio & Efficiency of Second System of Pulleys.

Apparatus: Second system of Pulley apparatus, small pan, threads or strings, slotted weights, small weight box, metre rod etc.

Theory: The arrangement of pulleys is as shown in fig 10.1(a). The apparatus consists of two 'blocks' of pulleys one of which is the fixed block while the other is the movable block. The pulleys in the fixed block are all fixed and all pulleys in the movable block are all movable. In most of the laboratories, the number of pulleys in both the blocks is same, however in some we may also have number of pulleys in fixed block one more than the number of pulleys in the movable block as shown in fig 10.1(b).

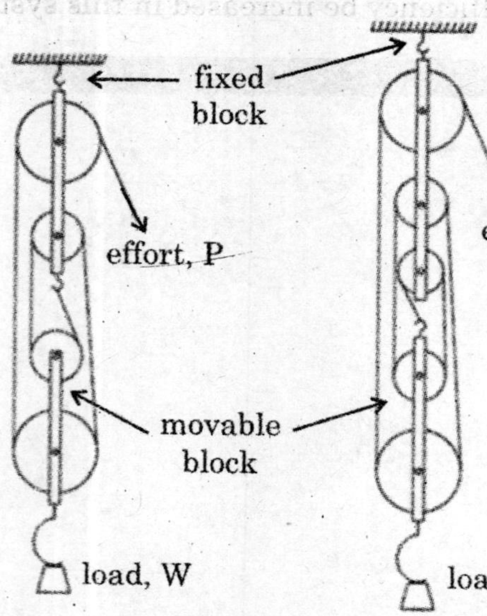

Fig. 10.1(a) & (b)

In this arrangement we have only one thread or rope that moves round the whole system. Starting from the end where the effort is applied, the rope goes round the top-most pulley in the fixed block then all the way down to the lower-most pulley in the movable block. The thread after going round this pulley is further taken round the next pulley in the fixed block and so on. The end of the rope is hooked to the fixed block if the number of pulleys is same in both blocks. This system is also known as 'pulley block system'.

Consider the equilibrium condition,

If the number of threads or ropes supporting the movable block is 'n' (4 in case of fig 10.1(a) & 5 in case of fig 10.1(b)) then,

$$W = n.P$$

Or,
$$\frac{W}{P} = n$$

Now,

a. Mechanical Advantage – It is defined as the ratio of Load lifted and the Effort applied.

$$M.A. = \frac{W}{P} = n$$

b. Velocity Ratio – It is defined as the ratio of the distance moved by the Effort and the Load. Here,

$$V.R. = \frac{\text{distance moved by effort}}{\text{distance moved by the load}} = n$$

c. Efficiency – It is defined as the ratio of the Mechanical Advantage and the Velocity Ratio. It is expressed as a percentage.

$$\eta = \frac{M.A.}{V.R.} \times 100$$

Note: The values derived above are theoretical and the effect of friction and weight of pulleys has been entirely neglected.

Procedure:

1. Make sure that the apparatus is properly functioning with minimum resistance or friction.
2. Hang some weight to the hook at the end of movable pulley block. Record it as Load, W.

3. Determine the weight of empty effort pan. Let this be 'w'. Add some weight to the effort pan. Keeping on adding more weight until the load W just starts lifting up. Record this value of Effort as 'P'.
4. Repeat the above steps 3 more times to obtain total 4 different sets of readings.

For Velocity Ratio:

1. Determine the initial position of the effort pan and the load with respect to the ground with the help of metre rod. Let these be 'y_1' and 'x_1' respectively.
2. Now, pull the effort pan down a certain distance and measure the height of the effort pan and the load with respect to ground. Let these be 'y_2' and 'x_2' respectively.

Observations:

Number of pulleys in the system, n = ...

Weight of the empty pan, w = ... g.

Initial position of effort pan, y_1 = ... cm.

Final position of effort pan, y_2 = ... cm.

Initial position of Load, x_1 = ... cm.

Final position of Load, x_2 = ... cm.

Table:

Serial No.	Load, W (g)	Effort, P (g) (including 'w')	M.A.	Theoretical M.A. = n	V.R.	Theoretical V.R. = n	Efficiency, η (%)
1.							
2.							
3.							

Calculations:

$$M.A. = \frac{W}{P}$$

$$\text{V.R.} = \frac{y_2 - y_1}{x_2 - x_1}$$

$$\text{Efficiency, } \eta = \frac{M.A.}{V.R.} \times 100$$

Results:

For the given 2nd system of pulley,

1. Mechanical Advantage = ...
2. Velocity ratio = ...
3. Efficiency = ... %.

Discussions:

1. What do you understand by movable and fixed pulleys?
2. Application of effort is easier in which type of pulley?
3. To what extent can the efficiency of this system of pulley be practically increased?

Space for Answers:

EXPERIMENT No. 11

To determine the Mechanical Advantage, Velocity Ratio & Efficiency of Third system of Pulleys

Lab Objectives: To determine the Mechanical Advantage, Velocity Ratio & Efficiency of Third System of Pulleys.

Apparatus: Third system of Pulley apparatus, small pan, threads or strings, slotted weights, small weight box, metre rod etc.

Theory: The arrangement of pulleys is as shown in fig 11.1. The apparatus generally consists of four pulleys out of which one is fixed while all others are movable. The first pulley from the top is fixed to a rigid bar and all the other movable pulleys are subsequently positioned below it as shown in fig 11.1.

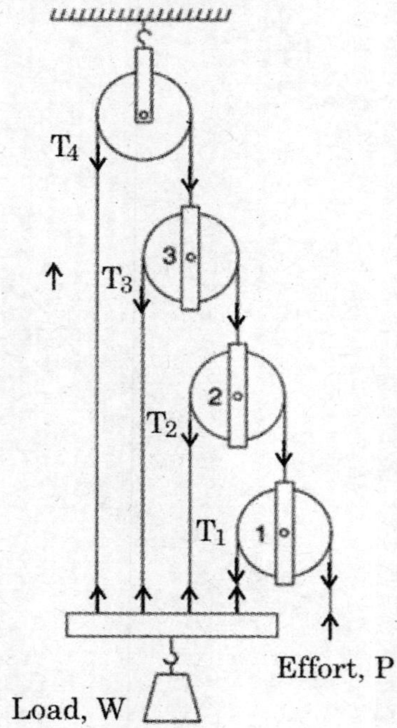

Fig. 11.1.

A separate string runs round each pulley. One end of each string is connected to the load to be lifted through a flat plate while the other end, after going round the pulley is hooked to the top of just lower pulley block. At the free end of the string going round the lowest pulley effort is applied. This system is also known as 'separate string attached to weight system'.

Consider the equilibrium condition,

Let 'n' be the number of pulleys in the system (including fixed pulley)

Here $n = 4$

$$W = T_1 + T_2 + T_3 + T_4$$

But, $T_1 = P$,

$$T_2 = 2T_1 = 2P,$$

$$T_3 = 2T_2 = 2^2 P$$

$$T_4 = 2T_3 = 2^3 P$$

$$\therefore W = P + 2^1 P + 2^2 P + 2^3 P$$

Or $W = P(1+2^1+2^2+2^3) = P(2^4-1)$ (sum of 4 terms of GP)

For n-terms, $W = P.\dfrac{2^n-1}{2-1}$

Or $\dfrac{W}{P} = \dfrac{2^n-1}{2-1} = 2^n - 1$

Now,

a. Mechanical Advantage – It is defined as the ratio of Load lifted and the Effort applied.

$$M.A. = \frac{W}{P} = 2^n - 1$$

b. Velocity Ratio – It is defined as the ratio of the distance moved by the Effort and the Load. Here,

$$V.R. = \frac{\text{distance moved by effort}}{\text{distance moved by the load}} = 2^n - 1$$

c. Efficiency – It is defined as the ratio of the Mechanical Advantage and the Velocity Ratio. It is expressed as a percentage.

$$\eta = \frac{M.A.}{V.R.} \times 100$$

Note: The values derived above are theoretical and the effect of friction and weight of pulleys has been entirely neglected.

Procedure:

1. Make sure that the apparatus is properly functioning with minimum resistance or friction.
2. Attach some weight to the hook fixed below the flat plate. Record it as Load, W.
3. Determine the weight of empty effort pan. Let this be ‘w’. Add some weight to the effort pan. Keeping on adding more weight until the load W just starts lifting up. Record this value of Effort as ‘P’.
4. Repeat the above two steps 3 more times to obtain total 4 different sets of readings.

For Velocity Ratio:

1. Determine the initial position of the effort pan and the load with respect to the ground with the help of metre rod. Let these be ‘y_1’ and ‘x_1’ respectively.
2. Now, pull the effort pan down a certain distance and measure the height of the effort pan and the load with respect to ground. Let these be ‘y_2’ and ‘x_2’ respectively.

Observations:

Number of pulleys in the system, n = ...

Weight of the empty pan, w = ... g.

Initial position of effort pan, y_1 = ... cm.

Final position of effort pan, y_2 = ... cm.

Initial position of Load, x_1 = ... cm.

Final position of Load, x_2 = ... cm.

Table:

Serial No.	Load, W (g)	Effort, P (g) (including 'w')	M.A.	Theoretical M.A. = $2^n - 1$	V.R.	Theoretical V.R. = $2^n - 1$	Efficiency, η (%)
1.							
2.							
3.							

Calculations:

$$\text{M.A.} = \frac{W}{P}$$

$$\text{V.R.} = \frac{y_2 - y_1}{x_2 - x_1}$$

$$\text{Efficiency, } \eta = \frac{M.A.}{V.R.} \times 100$$

Results:

For the given 3rd system of pulley,

1. Mechanical Advantage = ...
2. Velocity ratio = ...
3. Efficiency = ...%.

Discussions:

1. Out of the three systems of pulley which one is most efficient and why?
2. How can the efficiency of any system of pulley be increased? Does this have any relation to the mode of method of application of effort? Give reasons to support your view.

Space for Answers:

EXPERIMENT No. 12

To determine the Mechanical Advantage, Velocity Ratio, Efficiency& Law of Machine for Weston's Pulley block

Lab Objectives: To determine the Mechanical Advantage, Velocity Ratio, Efficiency & Law of Machine of Weston's Differential Pulley block.

Apparatus: Weston's differential pulley block apparatus, small pan, threads or strings, chain, slotted weights, small weight box, metre rod etc.

Theory: It is yet another pulley system that is employed for lifting heavy weights. It has a very peculiar design. The arrangement consists of total three pulleys. Two pulleys are keyed to a common spindle and are fixed to a support. These two pulleys have different diameter, one smaller than the other. The third pulley is basically a movable pulley which (as in most of the cases) is hooked to the load to be lifted.

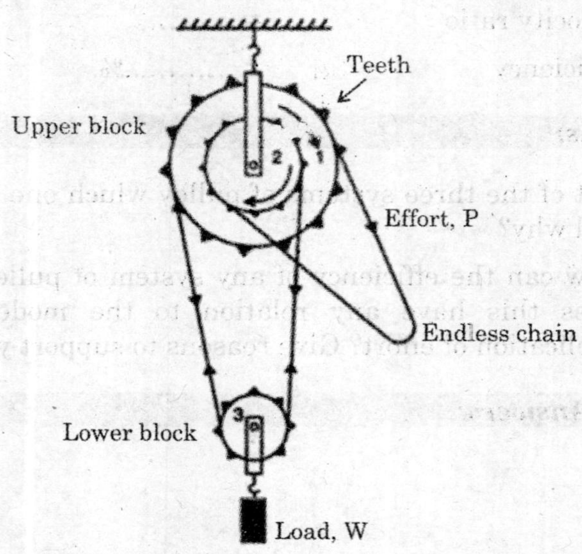

Fig. 12.1.

An endless chain runs round larger diameter pulley of the upper block and then after going round the bottom of the lower pulley is again taken to the upper block to go round the smaller diameter pulley of the upper block. Thus we observe that only one endless chain moves round all the pulleys. The pulleys used in the setup are toothed, so as to prevent slippage between chain and pulleys (See fig 12.1).

For better understanding of movement of endless chain, arrow has been marked in only one direction in fig 12.1.

Let,

D = diameter of the pulley 1.

d = diameter of the pulley 2.

W = load lifted.

P = effort applied.

For one revolution of the pulley 1, (refer fig 12.1)

Length of the chain pulled down = πD

Or, the distance moved by the effort = πD

Length of the chain pulled up (round the pulley 2) = πD

$\therefore$ Net shortening of the chain = $\pi D - \pi d = \pi (D - d)$

This net shortening is equally distributed between two branches of chain round the pulley 3.

Thus, distance through which the load is lifted = $\frac{1}{2}.\pi (D-d)$

Now,

a. Mechanical Advantage – It is defined as the ratio of Load lifted and the Effort applied.

$$M.A. = \frac{W}{P}$$

b. Velocity Ratio – It is defined as the ratio of the distance moved by the Effort and the Load. Here,

$$V.R. = \frac{\text{distance moved by effort}}{\text{distance moved by the load}}$$

$$= \frac{\pi D}{\frac{1}{2}.\pi (D-d)} = \frac{2D}{(D-d)}$$

c. Efficiency – It is defined as the ratio of the Mechanical Advantage and the Velocity Ratio. It is expressed as a percentage.

$$\eta = \frac{M.A.}{V.R.} \times 100$$

d. Law of Machine – It gives a relation between the Load lifted and the Effort applied. If a graph is plotted between 'W' & 'P' then it will be observed that it is a straight line of the form

$$P = mW + C$$

Clearly the straight line does not passes through origin but makes an intercept 'C' on the P-axis thus showing that some effort is used up to overcome friction before the load is lifted.

Procedure:

1. Make sure that the apparatus is properly functioning with minimum resistance or friction.
2. Hang some weight to the movable block though the hanger or the hook. Record it as Load, W.
3. Install the effort pan at least 30cm above the load.
4. Determine the weight of empty effort pan. Let this be 'w'. Add some weight to the effort pan. Keeping on adding more weight until the load. W just starts lifting. Record this value of Effort as 'P'.
5. Repeat the above three steps 2 more times to obtain total 3 different sets of readings.

For Velocity Ratio:

1. Determine the initial position of the effort pan and the load with respect to the ground with the help of metre rod. Let these be 'y_1' and 'x_1' respectively.
2. Now, pull down the effort pan down a certain distance and measure the height of the effort pan and the load with respect to ground. Let these be 'y_2' and 'x_2' respectively.

Observations:

Weight of the empty pan, w = ... g.

Initial position of effort pan, y_1 = ... cm.

Final position of effort pan, y_2 = ... cm.

Initial position of Load, x_1 = ... cm.

Final position of Load, x_2 = ... cm.

Table:

Serial No.	Load, W (g)	Effort, P (g) (including 'w')	M.A.	V.R.	Theoretical V.R. $= \frac{2D}{(D-d)}$	Efficiency, η (%)
1.						
2.						
3.						

Calculations:

$$\text{M.A.} = \frac{W}{P}$$

$$\text{V.R.} = \frac{y_2 - y_1}{x_2 - x_1}$$

$$\text{Efficiency, } \eta = \frac{M.A.}{V.R.} \times 100$$

To obtain the Law of Machine, take any two values of Load & corresponding Effort. Let these be W_1 & W_2 and P_1 & P_2.

Put these values one by one in the equation, $P = mW + C$ and solve the two equations to obtain values of 'm' and 'C'.

Substitute the values of 'm' and 'C' in $P = mW + C$ to get the Law of Machine.

Results:

For the given Weston's differential pulley block,

1. Mechanical Advantage = ...
2. Velocity ratio = ...
3. Efficiency = ...%.

Discussions:

1. Why do we call this system a 'differential pulley block system'?
2. What is the role of the lower pulley block?

Space for Answers:

EXPERIMENT No. 13

To determine the Mechanical Advantage, Velocity Ratio, Efficiency & Law of Machine for Worm and Worm Wheel

Lab Objective: To determine the Mechanical Advantage, Velocity Ratio, Efficiency & Law of Machine for Worm and Worm Wheel.

Apparatus: Worm and Worm Wheel apparatus, slotted weights, rope, small hooks, small pan, metre rod, Vernier Calipers etc.

Theory: Worm and Worm Wheel is basically a simple machine that serves the purpose of lifting heavy weights by application of small effort. The apparatus consist of two major components viz.

a Worm - It is fundamentally a square threaded screw with a single start. Screws with two or multiple start, add up to the mechanical advantage of the device.

b Worm Wheel – It is a toothed wheel on which is securely mounted the Load Drum.

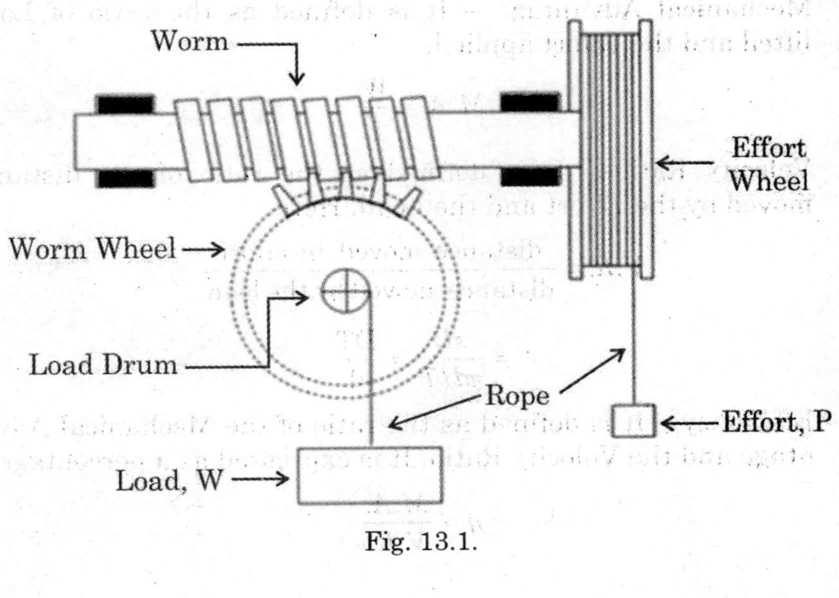

Fig. 13.1.

Worm and Worm Wheel are properly geared with each other. An effort wheel a pulley is attached to the end of the Worm. On this wheel (or pulley), a rope is wound and to its free end a small pan is attached, here the effort is applied. Sometimes, a handle is also provided that completely dispenses with the wheel (or pulley).

A rope is also wound round the Load Drum and to its free end a hook is tied. To this hook, the load to be lifted is attached (see fig 13.1).

Let,

T = Number of teeth on the Worm Wheel

D = Diameter of the Effort Wheel

d = Diameter of the Worm Wheel

P = Effort applied

W = Load lifted

For one revolution of Effort Wheel and hence the Worm, distance moved by the Effort is 'πD'. The Worm Wheel moves a tooth ahead or back (as the case may be).

Thus, the Load Drum undergoes $\frac{1}{T}$ revolution.

The distance through which load is lifted = $\frac{\pi d}{T}$

Now,

a. Mechanical Advantage – It is defined as the ratio of Load lifted and the Effort applied.

$$M.A. = \frac{W}{P}$$

b. Velocity Ratio – It is defined as the ratio of the distance moved by the Effort and the Load. Here,

$$V.R. = \frac{\text{distance moved by effort}}{\text{distance moved by the load}}$$

$$= \frac{\pi D}{\pi d / T} = \frac{\text{DT}}{\text{d}}$$

c. Efficiency – It is defined as the ratio of the Mechanical Advantage and the Velocity Ratio. It is expressed as a percentage.

$$\eta = \frac{M.A.}{V.R.}$$

$$= \frac{W}{P} \times \frac{d}{DT} \times 100$$

d. Law of Machine – It gives a relation between the Load lifted and the Effort applied. If a graph is plotted between 'W' & 'P' then it will be observed that it is a straight line of the form

$$P = mW + C$$

Clearly the straight line does not passes through origin but makes an intercept 'C' on the P-axis thus showing that some effort is used up to overcome friction before the load is lifted.

Procedure:

1. Measure and record the diameter of Effort Wheel and the Worm Wheel using the calipers.
2. Attach some load to the hook at the end of rope running round the Worm Wheel. Let this load be 'W'. Record its value.
3. Determine the weight of empty effort pan. Let this be 'w'.
4. Now, add some weight to the effort pan. Gradually increase the weight until the Load just starts lifting up. Let this value of Effort be 'P'.
5. Repeat the above steps two more times to obtain 3 different sets of observations.

Observations:

Weight of the empty effort pan, w = ... g.

Diameter of the Effort Wheel, D = ... cm.

Diameter of the Worm Wheel, d = ... cm.

Number of teeth on the Worm Wheel, T = ...

Table:

Serial No.	Load, W (g)	Effort, P (g) (including 'w')	M.A.	V.R.	Efficiency, η (%)
1.					
2.					
3.					

Calculations:

$$\text{M.A.} = \frac{W}{P}$$

$$\text{V.R.} = \frac{DT}{d}$$

$$\text{Efficiency, } \eta = \frac{W}{P} \times \frac{d}{DT} \times 100$$

To obtain the Law of Machine, take any two values of Load & corresponding Effort. Let these be W_1 & W_2 and P_1 & P_2.

Put these values one by one in the equation, $P = mW + C$ and solve the two equations to obtain values of 'm' and 'C'.

Substitute the values of 'm' and 'C' in $P = mW + C$ to get the Law of Machine.

Results:

For the given Worm & Worm Wheel apparatus,

1. Mechanical Advantage = ...
2. Velocity ratio = ...
3. Efficiency = ... %.
4. Law of Machine ....

Discussions:

1. What do you understand by a machine?
2. How many types of machines are there? Give brief description of each.

Space for Answers:

EXPERIMENT No. 14

To determine the mechanical advantage, velocity ratio and efficiency of a Differential Wheel and Axle

Lab Objectives: To determine the mechanical advantage, velocity ratio and efficiency of a Differential Wheel and Axle.

Apparatus: Differential wheel and axle apparatus, hanger weights, slotted weights, metre rod, threads or strings, rope, Vernier calipers etc.

Theory: Differential Wheel and Axle is basically an enhancement over the basic wheel and axle mechanism. It is aimed at yielding higher velocity ratio compared to its predecessors.

The apparatus is rigidly mounted on a wall. Here, on a common shaft we have two load axles of different diameters. The one with larger diameter is referred to here as A and the smaller one as B. Also a wheel is mounted on the very same shaft. Since the rolling friction is smaller than the sliding friction, the common shaft is mounted at the end on ball bearing arrangement (See fig 14.1).

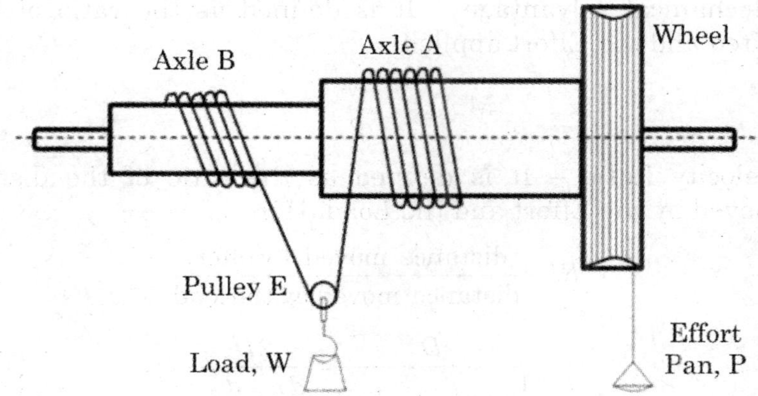

Fig. 14.1.

Now, a rope is hooked on the axle A and is wound a few turns round it. The same rope is attached to the hook and then taken round a pulley, E. This rope is further on taken to the axle B and is wound in the fashion opposite to that axle A and finally hooked on the same. Thus we see that a single rope runs round the axle A, the pulley E and the axle B. Another rope is taken and wound on the wheel in same fashion as on axle A. To the free end of this rope a small pan is provided for application of effort. The load to be lifted is hooked to the movable pulley, E.

Let,

D = diameter of wheel

d_A = diameter of axle A

d_B = diameter of axle B

W = load lifted

P = Effort applied

Consider one rotation of the wheel,

Distance moved by the effort = πD

Length of the rope unwound from axle A = πd_A

Length of the rope wounded on axle B = πd_B

Now the total length of rope wounded = $\pi d_B - \pi d_A$

Distance through which the pulley and thus the load moves

$$= \frac{1}{2} \times (\pi d_B - \pi d_A)$$

Now,

a. Mechanical Advantage – It is defined as the ratio of Load lifted and the Effort applied.

$$M.A. = \frac{W}{P}$$

b. Velocity Ratio – It is defined as the ratio of the distance moved by the Effort and the Load. Here,

$$V.R. = \frac{\text{distance moved by effort}}{\text{distance moved by the load}}$$

$$= \frac{\pi D}{\frac{1}{2} \times (\pi d_B - \pi d_A)} = \frac{2D}{d_B - d_A}$$

c. Efficiency – It is defined as the ratio of the Mechanical Advantage and the Velocity Ratio. It is expressed as a percentage.

$$\eta = \frac{M.A.}{V.R.}$$

$$= \frac{W}{P} \times \frac{d_B - d_A}{2D} \times 100$$

Procedure:

1. Measure and record the diameter of Wheel, the axle A & axle B using the calipers.
2. Attach some load to the hook fixed to the movable pulley E. Let this load be 'W'. Record its value.
3. Determine the weight of empty effort pan. Let this be 'w'.
4. Now, add some weight to the effort pan. Gradually increase the weight until the Load just starts lifting up. Let this value of Effort be 'P'.
5. Repeat the above steps two more times to obtain 3 different sets of observations.

Observations:

Weight of the empty effort pan, w = ... g.

Diameter of the Wheel, D = ... cm.

Diameter of the Axle A, d_A = ... cm.

Diameter of the Axle A, d_B = ... cm.

Table:

Serial No.	Load, W (g)	Effort, P (g) (including 'w')	M.A.	V.R.	Efficiency, η (%)
1.					
2.					
3.					

Calculations:

$$\text{M.A.} = \frac{W}{P}$$

$$\text{V.R.} = \frac{2D}{d_B - d_A}$$

$$\text{Efficiency, } \eta = \frac{W}{P} \times \frac{d_B - d_A}{2D} \times 100$$

Results:

For the given Differential Wheel & axle apparatus,

1. Mechanical Advantage = ...
2. Velocity ratio = ...
3. Efficiency = ... %.

Discussions:

1. Explain the basic wheel and axle mechanism.
2. How is Differential wheel & axle an enhancement over basic wheel and axle mechanism?

Space for Answers:

EXPERIMENT No. 15

To determine the Mechanical Advantage, Velocity Ratio, Efficiency & Law of Machine for a Screw Jack

Lab Objectives: To determine the Mechanical Advantage, Velocity Ratio, Efficiency & Law of Machine for a Screw Jack.

Apparatus: Screw jack apparatus, slotted weights, small pans, threads or strings, rope, metre rod, Vernier calipers etc.

Theory: Screw jack is a machine that is used for lifting heavy weights or loads by application of rather small effort. The machine is based on the Screw & nut principle and on principle of Inclined plane.

The Screw jack consists of a rod having square threads known as the screw and a threaded block known as the jack. The threads on the jack are made on its inner surface so that the threads on the outer surface of the screw properly grip into them (see fig 15.1).

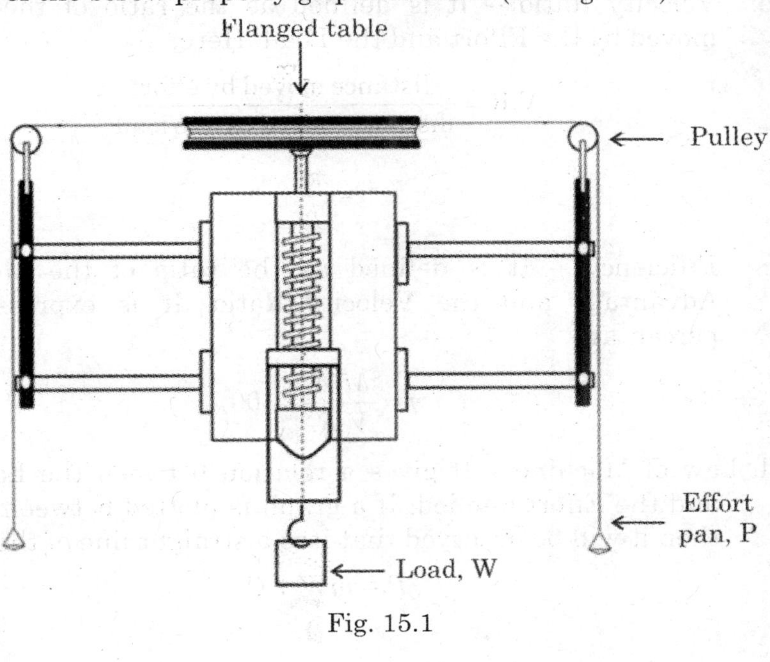

Fig. 15.1

Its mechanism is very simple. The load required to be lifted is placed on the broad head of the screw. However, in most of the laboratories we find that there is a different location provided for hooking of the load, but fundamentally the load lies on the same axis in both cases. The load is lifted by movement of the screw which is rotated by application of effort.

Let,

p = Pitch of the screw

d = diameter of the flanged table

W = Load to be lifted

P = Effort applied

Consider one revolution of the flanged table or the screw,

Distance through which the effort is moved = πd

Distance through which the load is lifted = 1 pitch = p

Now,

a. Mechanical Advantage – It is defined as the ratio of Load lifted and the Effort applied.

$$M.A. = \frac{W}{P}$$

b. Velocity Ratio – It is defined as the ratio of the distance moved by the Effort and the Load. Here,

$$V.R. = \frac{\text{distance moved by effort}}{\text{distance moved by the load}}$$

$$= \frac{\pi d}{p}$$

c. Efficiency – It is defined as the ratio of the Mechanical Advantage and the Velocity Ratio. It is expressed as a percentage.

$$\eta = \frac{M.A.}{V.R.} \times 100$$

d. Law of Machine – It gives a relation between the Load lifted and the Effort applied. If a graph is plotted between 'W' & 'P' then it will be observed that it is a straight line of the form

$$P = mW + C$$

Clearly the straight line does not passes through crigin but makes an intercept 'C' on the P-axis thus showing that some effort is used up to overcome friction before the load is lifted.

Procedure:

1. Wound a rope along the circumference of the flanged table and pass it over one of the pulley. To the free end of this rope fasten a small pan. Perform the similar action for another rope and make it pass over the other pulley.
2. Hang the load required to be lifted to the hook provided. Record it as 'W'.
3. Determine the weight of empty effort pan. Let this be 'w'.
4. Put some weight in the small effort pans. Keep on adding more weight until the load is lifted. Record this total value as 'P'.
5. Repeat the above procedure to obtain 3 different sets of reading.

For Velocity Ratio

1. Wrap a thread or a string round the circumference of the flanged table and determine its circumference using a metre rod.
2. Using Vernier Calipers measure the pitch of the screw.

Observations:

Weight of the empty pans, w = ... g.

Pitch of the screw, p = ...

Diameter of the flanged table, d = ... cm.

Table:

Serial No.	Load, W (g)	Effort, P (g) (including 'w')	M.A.	V.R.	Efficiency, η (%)
1.					
2.					
3.					

Calculations:

$$\text{M.A.} = \frac{W}{P}$$

$$\text{V.R.} = \frac{\pi d}{p}$$

$$\text{Efficiency, } \eta = \frac{W}{P} \times \frac{\pi d}{p} \times 100$$

To obtain the Law of Machine, take any two values of Load & corresponding Effort. Let these be W_1 & W_2 and P_1 & P_2.

Put these values one by one in the equation, $P = mW + C$ and solve the two simultaneous equations to obtain values of 'm' and 'C'.

Substitute the values of 'm' and 'C' in $P = mW + C$ to get the Law of Machine.

Results:

For the given Screw Jack apparatus,

1. Mechanical Advantage = ...
2. Velocity ratio = ...
3. Efficiency = ... %.
4. Law of Machine ....

Discussions:

1. Explain the theory behind Screw Jack apparatus.
2. Give some practical examples of a Screw Jack.

Space for Answers:

EXPERIMENT No. 16

To determine the Mechanical Advantage, Velocity Ratio & Efficiency of a Single Purchase Crab Winch

Lab Objectives : To determine the Mechanical Advantage, Velocity Ratio & Efficiency of a Single Purchase Crab Winch.

Apparatus : Single Purchase Crab Winch apparatus, weights, metre rod, small pan, ropes etc.

Theory : We know that, a simple machine consists of an axle, to which a wheel is fastened, so that torque applied to the wheel winds a rope or chain onto the axle, yielding a mechanical advantage. Single Purchase Crab Winch is the first advancement made to this simple wheel & axle mechanism.

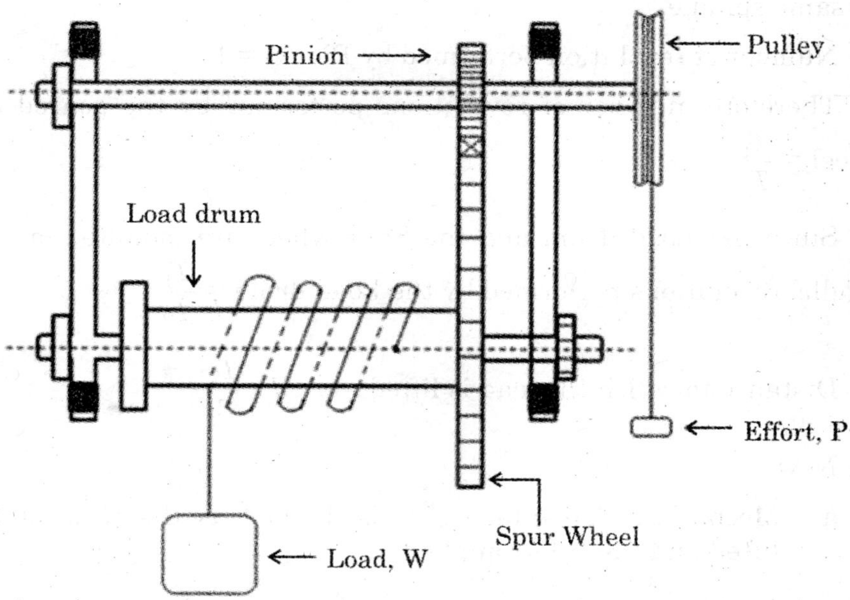

Fig. 16.1.

The system consists of two toothed wheels, one being much smaller than other one. The larger wheel is called the Spur Wheel and is mounted on a spindle or axle. On this same axle the Load drum is also mounted, by side of the Spur Wheel. On the Load drum a rope is hooked and is wound a few turn round it. The smaller toothed wheel is known as the Pinion. It is mounted on a separate axle, on which, at the end is provided a pulley. On this pulley is fixed rope which is also wound round it. To the end of this rope a small pan is attached for application of effort. The two toothed wheel are properly geared to each other (see fig 16.1).

Let,

T_1 = Number of teeth on the Pinion.

T_2 = Number of teeth on the Spur Wheel.

D = Diameter of the effort pulley.

d = Diameter of the Load drum.

W = Load to be lifted.

P = Effort applied.

For one revolution of Effort Wheel, distance moved by the Effort is 'πD'. Keeping in mind that, the Pinion and the pulley are mounted on the same spindle,

Number of revolution performed by Pinion = 1

Therefore, number of revolutions performed by the geared Spur Wheel = $\frac{T_1}{T_2}$

Since the Load drum and the Spur wheel are mounted on same spindle, revolutions performed by the Load drum = $\frac{T_1}{T_2}$

Distance to which the load is lifted = $\pi \times d \times \frac{T_1}{T_2}$

Now,

a. Mechanical Advantage – It is defined as the ratio of Load lifted and the Effort applied.

$$M.A. = \frac{W}{P}$$

b. Velocity Ratio – It is defined as the ratio of the distance moved by the Effort and the Load. Here,

$$V.R. = \frac{\text{distance moved by effort}}{\text{distance moved by the load}}$$

$$= \frac{\pi \times D \times T_2}{\pi \times d \times T_1} = \frac{DT_2}{dT_1}$$

c. Efficiency – It is defined as the ratio of the Mechanical Advantage and the Velocity Ratio. It is expressed as a percentage.

$$\eta = \frac{M.A.}{V.R.}$$

$$= \frac{W}{P} \times \frac{DT_2}{dT_1} \times 100$$

Procedure:

1. Measure and record the diameter of Effort Pulley and the Load drum using the calipers or indirectly, with the help of a thread and a metre rod (whichever is convenient).
2. Attach some load to the hook at the end of rope running round the Load drum. Let this load be 'W'. Record its value.
3. Determine the weight of empty effort pan. Let this be 'w'.
4. Now, add some weight to the effort pan. Gradually increase the weight until the Load just starts lifting up. Let this value of Effort be 'P'.
5. Repeat the above steps two more times to obtain 3 different sets of observations.

Observations:

Number of teeth on the Pinion, T_1 = ...

Number of teeth on the Spur Wheel, T_2 = ...

Diameter of the Load drum, d = ... cm.

Diameter of the Effort Pulley, D = ... cm.

Weight of the empty effort pan, w = ... g.

Table:

Serial No.	Load, W (g)	Effort, P (g) (including 'w')	M.A.	V.R.	Efficiency, η (%)
1.					
2.					
3.					

Calculations:

$$\text{M.A.} = \frac{W}{P}$$

$$\text{V.R.} = \frac{DT_2}{dT_1}$$

$$\text{Efficiency, } \eta = \frac{W}{P} \times \frac{DT_2}{dT_1} \times 100$$

Results:

For the given Single Purchase Crab Winch apparatus,

1. Mechanical Advantage = ...
2. Velocity ratio = ...
3. Efficiency = ... %.

Discussions:

1. Practically to what extent can the efficiency of Single Purchase Crab Winch be increased?
2. What are probable sources of error in this experiment?

Space for Answers:

EXPERIMENT No. 17

To determine the Mechanical Advantage, Velocity Ratio & Efficiency of a Double Purchase Crab Winch

Lab Objectives: To determine the Mechanical Advantage, Velocity Ratio, Efficiency and Law of machine of a Double Purchase Crab Winch.

Apparatus: Double Purchase Crab Winch apparatus, weights, ropes, small pan, metre rod, thread or strings etc.

Theory: The Double Purchase Winch Crab is an enhancement over the Single Purchase Crab Winch that is aimed at achieving greater velocity ratio than the latter.

It consists of two Pinions, PN_1 & PN_2 and two Spur Wheels, S_1 & S_2. Refer fig 17.1.

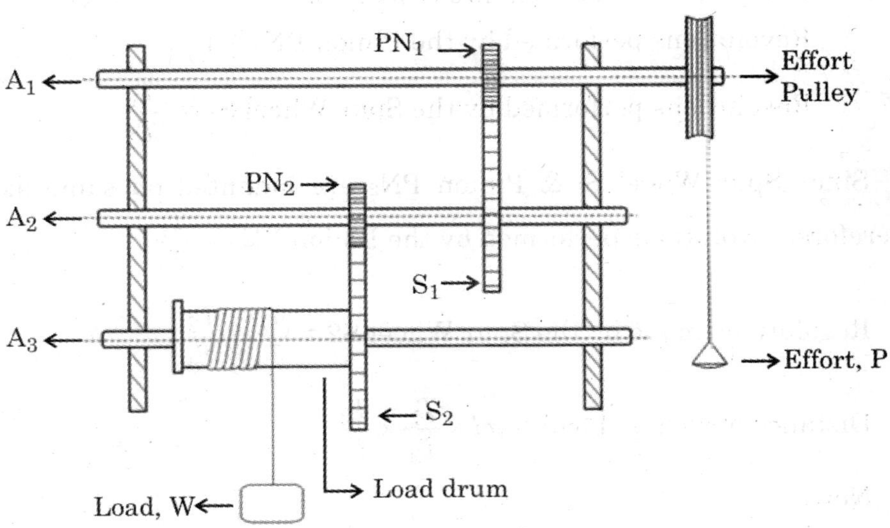

Fig. 17.1

We can clearly see in the fig 17.1, that the Pinion PN_1 meshes with Spur Wheel S_1. Also it can be seen that the Pinion PN_1 is mounted on the same axle A_1, on which the Effort Pulley is mounted.

Further, the Spur Wheel S_1 and the Pinion PN_2 are mounted on another axle A_2. Thus, the rotation of Spur Wheel S_1 is completely synchronised with the Pinion PN_2. The Pinion PN_2 meshes with the teeth of the Spur Wheel S_2, which is mounted on yet another axle A_3. Thus we have total 3 different axles in the whole setup.

The Load drum is also mounted on the axle A_3 and thus responds with the rotation of Spur Wheel S_2. On the Load drum is fixed a rope that is also wounded round it a few turns. At the end of this rope the required load to be lifted is hooked. The Effort is applied by adding weight to the effort pan that is suspended at the end of the rope wound round the Effort pulley.

Let,

T_1 = Number of teeth on Pinion PN_1.

T_2 = Number of teeth on Pinion PN_2.

T_3 = Number of teeth on Spur Wheel S_1.

T_4 = Number of teeth on Spur Wheel S_2.

D = Diameter of Effort pulley.

d = Diameter of Load drum.

Consider one revolution of the Effort pulley,

Distance moved by the Effort pan = πD

Revolutions performed by the Pinion PN_1 = 1

Revolutions performed by the Spur Wheel $S_1 = \frac{T_1}{T_3}$

Since Spur Wheel S_1 & Pinion PN_2 are mounted on same axle.

Therefore, revolutions performed by the Pinion $PN_2 = \frac{T_1}{T_3}$

Revolutions made by the Spur Wheel S2 = $\frac{T_1}{T_3} \times \frac{T_2}{T_4}$

Distance moved by Load = $\pi d \times \frac{T_1}{T_3} \times \frac{T_2}{T_4}$

Now,

a. Mechanical Advantage – It is defined as the ratio of Load lifted and the Effort applied.

$$M.A. = \frac{W}{P}$$

b. Velocity Ratio – It is defined as the ratio of the distance moved by the Effort and the Load. Here,

$$V.R. = \frac{\text{distance moved by effort}}{\text{distance moved by the load}}$$

$$= \frac{\pi D}{\pi d} \times \frac{T_3}{T_1} \times \frac{T_4}{T_2} = \frac{DT_3T_4}{dT_1T_2}$$

c. Efficiency – It is defined as the ratio of the Mechanical Advantage and the Velocity Ratio. It is expressed as a percentage.

$$\eta = \frac{M.A.}{V.R.}$$

$$= \frac{W}{P} \times \frac{DT_3T_4}{dT_1T_2} \times 100$$

d. Law of Machine – It gives a relation between the Load lifted and the Effort applied. If a graph is plotted between 'W' & 'P' then it will be observed that it is a straight line of the form

$$P = mW + C$$

Clearly the straight line does not passes through origin but makes an intercept 'C' on the P-axis thus showing that some effort is used up to overcome friction before the load is lifted.

Procedure:

1. Measure and record the diameter of Effort Pulley and the Load drum with the help of a thread and a metre rod.
2. Attach some load to the hook at the end of rope running round the Load drum. Let this load be 'W'. Record its value.
3. Determine the weight of empty effort pan. Let this be 'w'.
4. Now, add some weight to the effort pan. Gradually increase the weight until the Load just starts lifting up. Let this value of Effort be 'P'.
5. Repeat the above steps two more times to obtain 3 different sets of observations.

Observations:

Number of teeth on the Pinion PN_1, T_1 = ...

Number of teeth on the Pinion PN_2, T_2 = ...

Number of teeth on the Spur Wheel S_1, T_3 = ...

Number of teeth on the Spur Wheel S_2, T_4 = ...

Diameter of the Load drum, d = ... cm.

Diameter of the Effort Pulley, D = ... cm.

Weight of the empty effort pan, w = ... g.

Table:

Serial No.	Load, W (g)	Effort, P (g) (including 'w')	M.A.	V.R.	Efficiency, η (%)
1.					
2.					
3.					

Calculations:

$$M.A. = \frac{W}{P}$$

$$V.R. = \frac{DT_3\,T_4}{dT_1\,T_2}$$

$$\text{Efficiency, } \eta = \frac{W}{P} \times \frac{DT_3T_4}{dT_1T_2} \times 100$$

To obtain the Law of Machine, take any two values of Load & corresponding Effort. Let these be W_1 & W_2 and P_1 & P_2.

Put these values one by one in the equation, $P = mW + C$ and solve the two simultaneous equations to obtain values of 'm' and 'C'.

Substitute the values of 'm' and 'C' in $P = mW + C$ to get the Law of Machine.

Results:

For the given Double Purchase Crab Winch apparatus,

1. Mechanical Advantage = ...
2. Velocity ratio = ...
3. Efficiency = ... %.

4. Law of Machine ....

Discussions:

1. What is the basic difference between a Double Purchase Crab Winch & a Single Purchase Crab Winch?
2. Out of Single Purchase & Double Purchase Crab Winch which yields higher efficiency for given load? And why?

Space for Answers:

EXPERIMENT No. 18

To determine the mass moment of inertia of a Flywheel

Lab Objectives: To determine the mass moment of inertia of a Flywheel.

Apparatus: Flywheel, rope, a mat, metre rod, slotted weights with hangers, stop watch, threads or strings etc.

Theory: The flywheel is basically a big sized wheel. It is mounted on a metallic spindle and its circular surface is perpendicular to the longitudinal axis of this axle. This axle is supported at both the ends by a ball-bearing arrangement. On the axle a peg or a hook is provided. A thread having length slightly less than the height of the axle above the floor is fixed lightly to this peg. This thread is fixed in such a manner that it is just about to get detached from the peg. At the other end of this thread a suitable weight is hung through a hanger. Now slowly the axle is rotated so that the thread wounds itself on the circumference of the axle while taking care that there is no overlapping of loops occurs.

Clearly the system in this position is not in equilibrium. As soon as the weight is released and allowed to drop freely we observe

1. The potential energy of the weight decreases,
2. The axle and flywheel starts rotating. Thus flywheel gains rotational kinetic energy,
3. During its drop, the weight gains certain amount of kinetic energy,
4. Some energy is lost while overcoming friction at the supported ends of axle.

As the falling weight hits the mat on the floor, loosely attached thread gets detached.

Now, if

'r' is the radius of flywheel,

'm' is mass of the falling weight,

'v' is the average velocity of falling weight,

'h' is the height through which the weight falls,

'ω' is the angular velocity of rotation of the flywheel,

'I' is the moment of inertia of flywheel about its axis of rotation,

'W_f' is the work done against friction in one rotation of flywheel,

'n_1' is the number of rotations of flywheel when the weight hits the floor, then,

$$mgh = \frac{1}{2} I\omega^2 + \frac{1}{2} mv^2 + n_1 W_f \quad \text{...eq. 1}$$

Let 'n_2' be the number of rotations performed by the flywheel after the weight gets detached from it. Now the work done by the friction during these 'n_2' rotations is expended in bringing the flywheel to rest. Thus,

$$\frac{1}{2} I\omega^2 = n_2 W_f \quad \text{...eq. 2}$$

Or $$W_f = \frac{1 I\omega^2}{2 n_2} \quad \text{...eq. 3}$$

Let 't' be the time taken by the flywheel to come to rest after the detachment of weight.

Average angular velocity, $\overline{\omega} = \frac{2\pi n_2}{t}$

But average angular velocity, $\overline{\omega} = \frac{\omega_{initital} + \omega_{final}}{2}$

Here, $\omega_{initial} = \omega$; and $\omega_{final} = 0$

$$\therefore \overline{\omega} = \frac{\omega}{2}$$

Or $$\omega = 2\overline{\omega} = \frac{4\pi n_2}{t} \quad \text{...eq. 4}$$

Also, $v = \omega r$ ...eq. 5

Now, using eq. 3, eq. 4 and eq. 5 in eq. 1, we get,

$$I = \frac{(2mgh - mw^2 r^2)}{w^2 (1 + \frac{n_1}{n_2})}$$

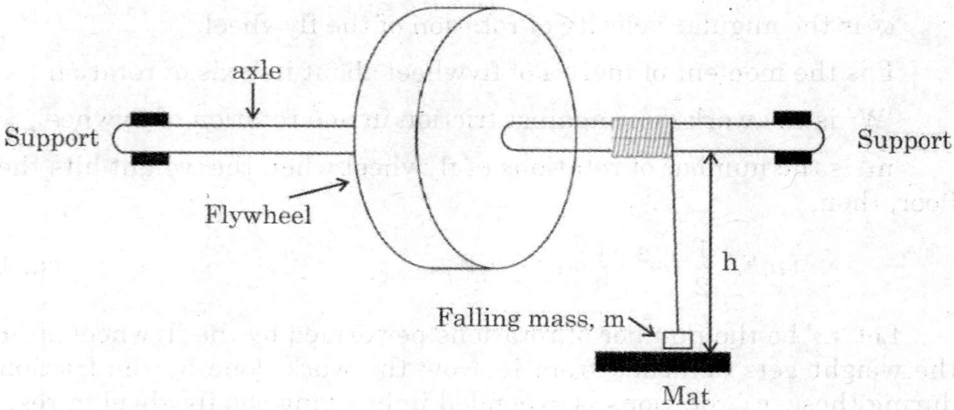

Fig. 18.1.

Procedure:

1. Measure the radius of the flywheel directly using calipers or indirectly after determining its circumference with thread and a metre rod.
2. Lightly fix one end of the rope to the peg on the axle. To the other end attach a mass of about say 500g. Record it as 'm'. The thread must be long enough to be just touching the mat on the floor.
3. Wound the thread round the axle by gently and slowly rotating the flywheel. Count the number of rotations of the flywheel. Record the number of rotations as 'n_1'.
4. Keep the mass and the flywheel in static condition and in this state itself, measure the height of the bottom surface of the mass from the mat. Record this height as 'h'.
5. Now, release the mass and start the stopwatch as soon as the mass strikes the mat and also start counting the rotations of the flywheel from then.
6. Record the time taken by the by the flywheel to come to a stop and record this time as 't'. The number of rotation performed by the flywheel is recorded as 'n_2'.
7. Repeat the above procedure with a different value of 'm'. Take in all three sets of readings.

Observations:

Radius of the flywheel, r = ... cm.

Acceleration due to gravity, g = ... cm²/s.

Table:

Serial No.	Falling mass, m (g)	Height of mass above mat, h (cm)	Number of rotations, n_1	Number of rotations, n_2	Time taken by flywheel to stop, t (s)	Angular velocity, ω (s^{-1})	Moment of Inertia, I (g-cm^2)
1.							
2.							
3.							

Calculations:

Angular velocity, $\omega = \dfrac{4\pi n_2}{t}$

Moment of Inertia, $I = \dfrac{(2mgh - mw^2 r^2)}{w^2 (1 + \dfrac{n_1}{n_2})}$

Calculate the average value of Moment of Inertia of flywheel.

Results:

1. The average value of Moment of Inertia of given flywheel is ....

Discussions:

1. What do you understand by Moment of Inertia?
2. Why is it required to determine the moment of inertia of flywheel at all?

Space for Answers:

Multiple Choice Questions

1. According to principle of conservation of energy, the total momentum of a system of masses in any direction remains constant unless acted upon by an external force in that direction.

   (a) True (b) False

2. The friction experienced by a body, when in motion, is known as

   (a) Rolling Friction (c) Dynamic Friction
   (b) Static Friction (d) Limiting Friction

3. Two balls of equal mass and of perfectly elastic material are lying on the floor. One of the ball with velocity v is made to strike the second ball. Both the balls after impact will move with a velocity

   (a) v (c) v/2
   (b) v/4 (d) v/8

4. The term 'force' may be defined as an agent which produces or tends to produce, destroys or tends to destroy motion.

   (a) True (b) False

5. The coefficient of restitution for elastic bodies is one.

   (a) True (b) False

6. A smooth cylinder lying on its convex surface remains in __________ equilibrium.

   (a) Stable (b) Unstable
   (c) Neutral

7. Coefficient of friction is the ratio of the limiting friction to the normal reaction between the two bodies.

   (a) True (b) False

8. Moment of inertia of a circular section about an axis perpendicular to the section is

(a) $\pi d^3/16$ (c) $\pi d^3/32$

(b) $\pi d^4/32$ (d) $\pi d^4/64$

9. The unit of angular acceleration is

(a) Ns/m^2 (c) N-m

(b) m/s^2 (d) rad/s^2

10. The velocity ratio in case of an inclined plane inclined at angle θ to the horizontal and weight being pulled up the inclined plane by vertical effort is

(a) Sin θ (c) Cos θ

(b) Cosec θ (d) tan θ

11. A body of weight W is required to move up on rough inclined plane whose angle of inclination with the horizontal is α. The effort applied parallel to the plane is given by (where $\mu = \tan \varphi$ = Coefficient of friction between the plane and the body.)

(a) $P = W \tan\alpha$ (c) $P = W \tan(\alpha + \varphi)$

(b) $P = W (\sin\alpha + \mu \cos\alpha)$ (d) $P = W (\cos\alpha + \mu \sin\alpha)$

12. If the resultant of two equal forces has the same magnitude as either of the forces, then the angle between the two forces is

(a) 30° (c) 60°

(b) 90° (d) 120°

13. The force required to move the body up the plane will be minimum if it makes an angle with the inclined plane __________ the angle of friction.

(a) Equal to (c) Less than

(b) Greater than (d) Cannot be said

14. In the shown figure, if the angle of inclination of the plane is increased, then acceleration of the system will

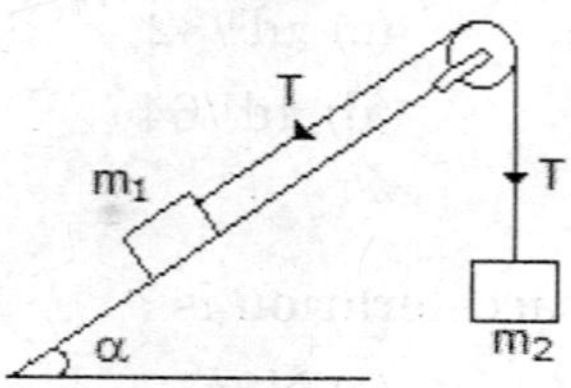

(a) Increase (c) Decrease

(b) Remains Same (d) Insufficient Data

15. The centre of gravity of an equilateral triangle with each side a, is __________ from any of the three sides.

(a) $\sqrt{3}\,a/2$ (c) $2\sqrt{3}\,a$

(b) $\dfrac{a}{2\sqrt{3}}$ (d) $3\sqrt{2}\,a$

16. The algebraic sum of the resolved parts of a number of forces in a given direction is equal to the resolved part of their resultant in the same direction. This is known as

(a) principle of independence of forces

(b) principle of transmissibility of forces

(c) principle of resolution of forces

(d) None of these

17. The point, through which the whole weight of the body acts, irrespective of its position, is known as

(a) Moment of Inertia (c) Centre of mass

(b) Centre of Gravity (d) Centre of Percussion

18. Which of the following is a scalar quantity?

(a) Pressure (c) Speed

(b) Acceleration (d) Force

19. The rate of change of momentum is directly proportional to the impressed force, and takes place in the same direction in which the force acts. This statement is known as

(a) Newton's 1st Law of motion (c) Newton's 2nd Law of motion

(b) Newton's 3rd Law of Motion (d) None of the above

20. The triangle law of forces states that if two forces acting simultaneously on a particle, be represented in magnitude and direction by the two sides of a triangle taken in order, then their resultant may be represented in magnitude and direction by the third side of a triangle, taken in opposite order.

(a) True (b) False

21. A lift moves downwards with an acceleration of 9.8 m/s^2. The pressure exerted by a man on the floor of the lift is zero.

(a) True (b) False

22. Non-coplanar non-concurrent forces are those forces which

(a) meet at one point, but their lines of action do not lie on the same plane

(b) do not meet at one point and their lines of action do not lie on the same plane

(c) meet at one point and their lines of action also lie on the same plane

(d) do not meet at one point, but their lines of action lie on the same plane

23. During elastic impact, the relative velocity of the two bodies after impact is __________ the relative velocity of the two bodies before impact.

(a) Equal to (c) Equal to & opposite

(b) Less than (d) Greater than

24. The skidding away of the vehicle on a level circular path can be avoided if the force of friction between the wheels and the ground is __________ the centrifugal force.

(a) Less than (b) Greater than

25. The motion of the body from one extremity to the other is known as a beat.

(a) True (b) False

26. A framed structure is perfect, if the number of members are __________ $(2j - 3)$, where j is the number of joints.

(a) Equal to (c) Less than

(b) Greater than (d) Either (b) or (c)

27. The maximum efficiency of a lifting machine is

(a) m/(VR) (c) 1/m

(b) 1/(VmR) (d) VR/m

28. A differential pulley block has larger and smaller diameters of 100 mm and 80 mm respectively. Its velocity ratio is

(a) 10 (c) 20

(b) 40 (d) 5

29. In ideal machines, mechanical advantage is _________ velocity ratio.

(a) Equal to (c) Greater than

(b) Less Than (d) None of these

30. All the steel trusses of the bridges, have one of their end roller supported, and other end hinged. The main advantage of such a support is that the truss remains stable.

(a) True (b) False

31. A machine having an efficiency less than 50%, is known as

(a) Neither reversible nor non-reversible Machine

(b) Non-reversible Machine

(c) Reversible Machine

(d) Ideal Machine

32. If the masses of both the bodies, as shown in the below figure, are doubled, then the acceleration in the string will be

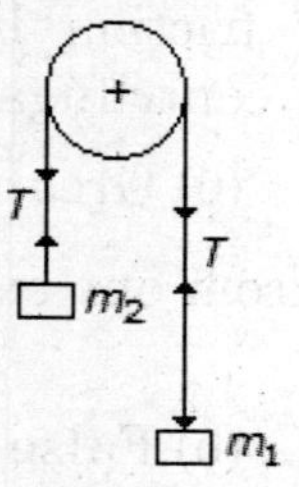

(a) Same (c) Double

(b) Half (d) Cannot be said

33. The centre of gravity of a hemisphere lies at a distance of $3r / 8$ from its base measured along the vertical radius.

(a) True b) False

34. The figure shows the three coplanar forces P, Q and R acting at a point O. If these forces are in equilibrium, then

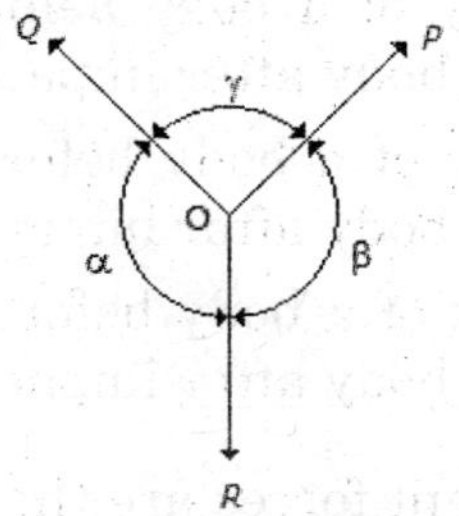

(a) $\frac{P}{\sin\beta}=\frac{Q}{\sin\alpha}=\frac{R}{\sin\gamma}$ (c) $\frac{P}{\sin\alpha}=\frac{Q}{\sin\beta}=\frac{R}{\sin\gamma}$

(b) $\frac{P}{\sin\gamma}=\frac{Q}{\sin\alpha}=\frac{R}{\sin\beta}$ (d) $\frac{P}{\sin\alpha}=\frac{Q}{\sin\gamma}=\frac{R}{\sin\beta}$

35. The force induced in the string AB due to the load W as shown in the below figure is

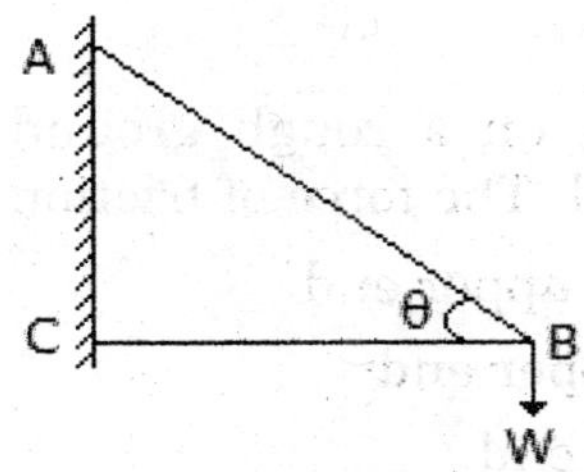

(a) $W\sin\theta$ (c) $W\cos\theta$

(b) $W\tan\theta$ (d) $W\operatorname{cosec}\theta$

36. The motion of a particle round a fixed axis is

(a) Rotary (c) Translatory

(b) Circular (d) Translatory as well as Rotary

37. The velocity ratio of a differential wheel and axle with D as the diameter of effort wheel and d_1, and d_2 as the diameters of larger and smaller axles respectively, is

(a) $\frac{D}{d_1+d_2}$ (c) $\frac{D}{d_1-d_2}$

(b) $\frac{2D}{d_1+d_2}$ (d) $\frac{2D}{d_1-d_2}$

38. Which of the following statement is correct?

(a) The kinetic energy of a body during impact remains constant.

(b) The kinetic energy of a body before impact is equal to the kinetic energy of a body after impact.

(c) The kinetic energy of a body before impact is less than the kinetic energy of a body after impact.

(d) The kinetic energy of a body before impact is more than the kinetic energy of a body after impact.

39. Non-coplanar concurrent forces are those forces which

(a) Do not meet at one point and their lines of action do not lie on the same plane.

(b) Meet at one point, but their lines of action do not lie on the same plane.

(c) Meet at one point and their lines of action also lie on the same plane.

(d) Do not meet at one point, but their lines of action lie on the same plane.

40. A ladder is resting on a rough ground and leaning against a smooth vertical wall. The force of friction will act

(a) downward at its upper end

(b) upward at its upper end

(c) zero at its upper end

(d) perpendicular to the wall at its upper end

41. A smooth cylinder lying on a __________ is in neutral equilibrium.

(a) Curved surface (c) Convex Surface

(b) Horizontal Surface (d) Cannot be said

42. Varingon's theorem of moments states that if a number of coplanar forces acting on a particle are in equilibrium, then

(a) their algebraic sum is zero

(b) their lines of action are at equal distances

(c) the algebraic sum of their moments about any point in their plane is zero

(d) the algebraic sum of their moments about any point is equal to the moment of their resultant force about the same point.

43. A spherical body is symmetrical about its perpendicular axis. According to Routh's rule, the moment of inertia of a body about an axis passing through its centre of gravity is (where, M = Mass of the body, and S = Sum of the squares of the two semi-axes.)

(a) MS/3

(c) MS/4

(b) MS/5

(d) None of the above

44. The minimum force required to slide a body of weight 'W' on a rough horizontal plane is

(a) W cos θ

(c) W sin θ

(b) W tan θ

(d) none of these

45. The velocity ratio in case of an inclined plane inclined at angle θ to the horizontal and weight being pulled up the inclined plane by vertical effort is

(a) Cos θ

(c) Sin θ

(b) tan θ

(d) none of the above

46. The efficiency of a lifting machine is the ratio of

(a) output to the input

(b) work done by the machine to the work done on the machine

(c) mechanical advantage to the velocity ratio

(d) all of the above

47. In actual machines, mechanical advantage is _________ velocity ratio.

(a) Equal to

(c) Less than

(b) Greater than

(d) Depends on machine

48. In a framed structure, as shown in the below figure, the forces in the members AB and BC are respectively

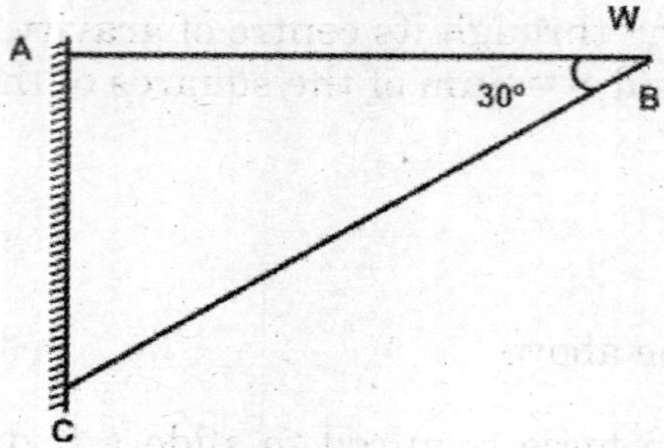

(a) $\sqrt{3}W$ (tensile) and $2W$ (compressive)

(b) $2W$ (tensile) and $\sqrt{3}W$ (compressive)

(c) $2\sqrt{3}W$ (tensile) and $2\sqrt{3}W$ (compressive)

(d) None of the above

49. The unit of moment of inertia of an area is

(a) Kg-m^2 (c) kg-m-s^2

(b) Kg/m^2 (d) m^4

50. The velocity ratio of a single purchase crab winch can be increased by

(a) increasing the length of the handle

(b) increasing the radius of the load drum

(c) increasing the number of teeth of the pinion

(d) all of the above

51. The terms 'leverage' and 'mechanical advantage' of a compound lever have got the same meaning.

(a) True (b) False

52. A lifting machine lifts a load of 1000N through a distance of 0.2 m by means of an effort of 200N through a distance of 1m. This machine is an ideal one.

(a) True (b) False

53. The angle which the normal reaction makes with the resultant reaction is called angle of friction.

(a) True (b) False

54. The moment of the force P about O as shown in the below figure is

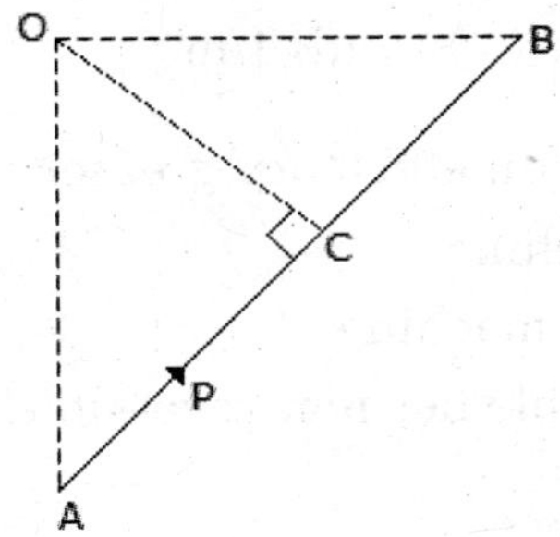

(a) P x OA (c) P x OB
(b) P x OC (d) P x AC

55. According to lami's theorem
(a) the three forces must be equal
(b) the three forces must be at 120° to each other
(c) the three forces must be in equilibrium
(d) if the three forces acting at a point are in equilibrium, then each force is proportional to the sine of the angle between the other two

56. When a body falls freely under gravitational force, it possesses __________ weight.
(a) No (c) Minimum
(b) Maximum (d) Cannot be said

57. In a screw jack, the effort required to lower the load is __________ the effort required to raise the same load.
(a) Less than (c) Equal to
(b) Greater than (d) None of the above

58. The static friction
(a) bears a constant ratio to the normal reaction between the two surfaces
(b) is independent of the area of contact, between the two surfaces
(c) always acts in a direction, opposite to that in which the body tends to move
(d) all of the above

59. The maximum mechanical advantage of a lifting machine is

(a) 1 + m (c) 1 – m

(b) m (d) 1/m

60. A machine having an efficiency greater than 50%, is known as

(a) Reversible machine

(b) Non-reversible machine

(c) Neither reversible nor non-reversible machine

(d) Ideal machine

61. For a self-locking machine, the efficiency must be

(a) Equal to 50% (b) Less than 50%

(c) Greater than 50% (d) 100%

62. The resultant of the two forces *P* and *Q* is R. If *Q* is doubled, the new resultant is perpendicular to *P*. Then

(a) P = Q (c) Q = R

(b) Q = 2R (d) None of these

63. Moment of inertia of a triangular section of base (*b*) and height (*h*) about an axis passing through its vertex and parallel to the base, is ________ than that passing through its C.G. and parallel to the base.

(a) 9 times (c) 6 times

(b) 4 times (d) 2 times

64. The efficiency of a screw jack is maximum, when (where α = Helix angle, and φ = Angle of friction).

(a) $\alpha = 45° + \varphi/2$ (c) $\alpha = 45° - \varphi/2$

(b) $\alpha = 90° + \varphi$ (d) $\alpha = 90° - \varphi$

65. When the lift is moving upwards with some acceleration, the pressure exerted by a man is ________ proportional to its acceleration.

(a) Directly (b) Inversely

66. Efficiency of a screw jack is given by (where, α = Helix angle, and φ = Angle of friction).

(a) $\frac{\tan(\alpha+\phi)}{\tan\alpha}$ (c) $\frac{\tan(\alpha-\phi)}{\tan\alpha}$

(b) $\frac{\tan\alpha}{\tan(\alpha+\phi)}$ (d) $\frac{\tan\alpha}{\tan(\alpha+\phi)}$

67. The acceleration of a body sliding down an inclined surface is

(a) g sin θ (c) g cos θ

(b) g tan θ (d) none of these

68. If a body is acted upon by a number of coplanar non-concurrent forces, it may

(a) rotate about itself without moving

(b) move in any one direction rotating about itself

(c) be completely at rest

(d) all of these

69. The velocity ratio for the third system of pulleys is

(a) doubled (c) quadrupled

(b) remains same (d) half

70. If the gravitational acceleration at any place is doubled, then the weight of a body will be

(a) doubled (c) quadrupled

(b) remains same (d) half

71. The moment of inertia of a thin spherical shell of mass m and radius r, about its diameter is

(a) $mr^2/3$ (c) $2mr^2/3$

(b) $2mr^2/3$ (d) $3mr^2/5$

72. The law of motion involved in the recoil of gun is Newton's

(a) 1st law of motion (c) 2nd law of motion

(b) 3rd law of motion (d) none of these

73. A machine which is not capable of doing any work in the reversed direction, after the effort is removed, is called a reversible machine.

(a) True (b) False

74. If the number of pulleys in a system is equal to its velocity ratio, then it is a __________ system of pulleys

(a) 1st (c) 2nd

(b) 3rd (d) Cannot be said

75. A circular hole of 50 mm diameter is cut out from a circular disc of 100 mm diameter as shown in the below figure. The centre of gravity of the section will lie

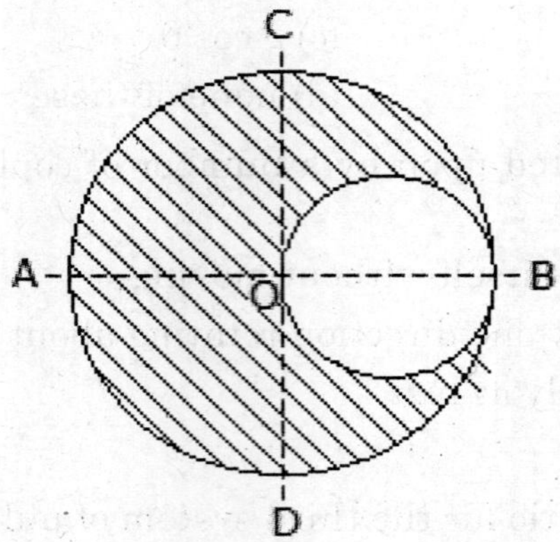

(a) In the shaded area (c) In the hole

(b) At O (d) Insufficient data

76. According to parallel axis theorem, the moment of inertia of a section about an axis parallel to the axis through centre of gravity (i.e. I_P) is given by (where, A = Area of the section, I_G = Moment of inertia of the section about an axis passing through its C.G., and h = Distance between C.G. and the parallel axis).

(a) $I_P = I_G + Ah^2$ (c) $I_P = I_G - Ah^2$

(b) $I_P = I_G / Ah^2$ (d) $I_P = Ah^2 / I_G$

77. The resultant of two equal forces P making an angle θ, is given by

(a) $2P \sin \theta /2$ (c) $2P \cos \theta /2$

(b) $2P \tan \theta /2$ (d) $2P \cot \theta /2$

78. The velocity ratio of a first system of pulleys with 4 pulleys is

(a) 4 (c) 8

(b) 16 (d) 20

79. A block of mass m_1, placed on an inclined smooth plane is connected by a light string passing over a smooth pulley to mass m_2, which moves vertically downwards as shown in the below figure. The tension in the string is

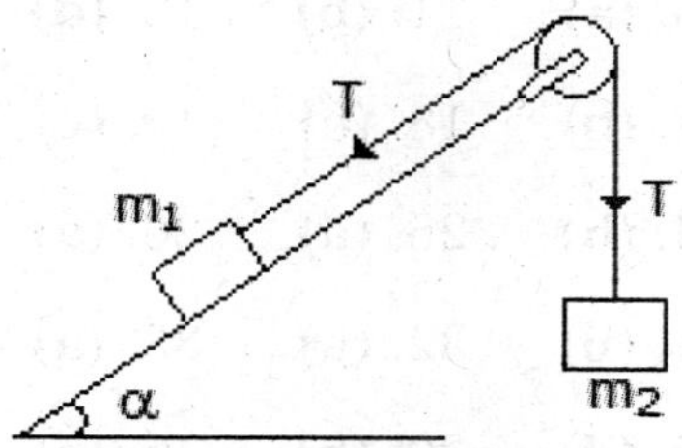

(a) m_1/m_2

(b) $m_1.g \sin \alpha$

(c) $m_1.m_2/m_1+m_2$

(d) $\frac{m_1 - m_2 . g(1+\sin\alpha)}{m_1+m_2}$

80. The maximum efficiency of a screw jack is

(a) $\frac{1-\sin\phi}{1+\sin\phi}$

(b) $\frac{1+\sin\phi}{1-\sin\phi}$

(c) $\frac{1-\tan\phi}{1+\tan\phi}$

(d) $\frac{1+\tan\phi}{1-\tan\phi}$

Answersheet

1. **(b)**	2. **(c)**	3. **(c)**	4. **(a)**	5. **(b)**	6. **(b)**	7. **(a)**
8. **(b)**	9. **(d)**	10. **(a)**	11. **(b)**	12. **(d)**	13. **(a)**	14. **(c)**
15. **(b)**	16. **(c)**	17. **(b)**	18. **(c)**	19. **(c)**	20. **(a)**	21. **(a)**
22. **(b)**	23. **(c)**	24. **(b)**	25. **(a)**	26. **(a)**	27. **(b)**	28. **(a)**
29. **(a)**	30. **(a)**	31. **(b)**	32. **(a)**	33. **(a)**	34. **(c)**	35. **(d)**
36. **(b)**	37. **(d)**	38. **(d)**	39. **(b)**	40. **(c)**	41. **(b)**	42. **(d)**
43. **(b)**	44. **(b)**	45. **(c)**	46. **(d)**	47. **(c)**	48. **(a)**	49. **(a)**
50. **(a)**	51. **(a)**	52. **(a)**	53. **(a)**	54. **(b)**	55. **(d)**	56. **(a)**
57. **(a)**	58. **(d)**	59. **(d)**	60. **(a)**	61. **(b)**	62. **(c)**	63. **(a)**
64. **(a)**	65. **(a)**	66. **(b)**	67. **(c)**	68. **(d)**	69. **(b)**	70. **(a)**
71. **(c)**	72. **(b)**	73. **(b)**	74. **(c)**	75. **(a)**	76. **(a)**	77. **(b)**
78. **(b)**	79. **(d)**	80. **(a)**				

Suggested Books

1. Engineering Mechanics by S.S. Bhavikatti, published by New Age International, 4th Edition, 2012.
2. Engineering Mechanics by Stephen Timoshenko, J.V. Rao, D.H. Young, published by Tata McGraw Hill Education, 4th Edition, 2006.
3. A Text Book of Applied Mechanics by R.K. Rajput, published by Laxmi Publications (P) Ltd, 3rd Edition, 2011.
4. Engineering mechanics by S C Mathur published by S K Kataria & Sons, First Edition 1998.
5. Engineering Mechanics by Dr I.S. Gujral, published by Laxmi Publications, First Edition, 2008.
6. Mechanics of Materials by B.C. Punmia, Laxmi Publications, First Edition 2001.
7. A Text Book of Engineering Mechanics by Dr R K Bansal, published by Laxmi Publications (P) Ltd., Sixth Edition, 2013.
8. Engineering Mechanics by D P Sharma, published by Pearson Education India, 1st Edition, 2011